Lecture Notes in Bioinformatics

Edited by S. Istrail, P. Pevzner, and M. Waterman

Subseries of Lecture Notes in Computer Science

Glenn Tesler   Dannie Durand (Eds.)

Comparative Genomics

International Workshop, RECOMB-CG 2007
San Diego, CA, USA, September 16-18, 2007
Proceedings

 Springer

Series Editors

Sorin Istrail, Brown University, Providence, RI, USA
Pavel Pevzner, University of California, San Diego, CA, USA
Michael Waterman, University of Southern California, Los Angeles, CA, USA

Volume Editors

Glenn Tesler
University of California, San Diego
Department of Mathematics
9500 Gilman Drive, La Jolla California 92093-0112, USA
E-mail: gptesler@math.ucsd.edu

Dannie Durand
Carnegie Mellon University
Departments of Biological Sciences and Computer Science
Pittsburgh, PA 15213, USA
E-mail: durand@cmu.edu

Library of Congress Control Number: 2007934768

CR Subject Classification (1998): F.2, G.3, E.1, H.2.8, J.3

LNCS Sublibrary: SL 8 – Bioinformatics

ISSN   0302-9743
ISBN-10   3-540-74959-4 Springer Berlin Heidelberg New York
ISBN-13   978-3-540-74959-2 Springer Berlin Heidelberg New York

Springer is a part of Springer Science+Business Media

springer.com

© Springer-Verlag Berlin Heidelberg 2007
Printed in Germany

Typesetting: Camera-ready by author, data conversion by Scientific Publishing Services, Chennai, India
Printed on acid-free paper   SPIN: 12124237   06/3180   5 4 3 2 1 0

Preface

The wealth of genomic data available today is a potential goldmine for basic research and economic development in the biomedical sciences. Comparison of related genomes offers enormous inferential power, revealing a wealth of knowledge about genome evolution, genetic function, and cellular processes. Recognition of this fact has spurred efforts to sequence a range of closely related primate and mammalian genomes, as well as concerted efforts to sequence multiple genomes in the yeast, *Drosophila*, and nematode lineages. Computational strategies to interpret and exploit these data are essential in order to realize the full value of these growing scientific resource. The annual RECOMB Satellite Workshop on Comparative Genomics (RECOMB-CG) is an interdisciplinary forum on all aspects of genome comparison, ranging from quantitative discoveries about genome structure to algorithms for comparative inference to theorems on the complexity of computational problems required for genome comparison.

This volume contains the papers presented at the Fifth Annual RECOMB Satellite Workshop on Comparative Genomics held September 16–18, 2007 in La Jolla, at the University of California, San Diego. Eighteen papers were submitted, of which the Program Committee selected 14 for presentation at the meeting and inclusion in these proceedings. Each submission was reviewed by at least three members of the Program Committee. The program also included a lively poster session. A session of short talks presenting late-breaking results, selected a few weeks before the meeting from the submitted poster abstracts, provided an opportunity to hear about provocative results from works in progress. In addition to contributed presentations, we were honored by plenary talks given by the invited speakers: Francesca Ciccarelli (European Institute of Oncology), Michael B. Eisen (University of California, Berkeley), Matthew Hahn (Indiana University), Katherine S. Pollard (University of California, Davis), Oliver A. Ryder (Zoological Society of San Diego), and Ajit Varki (University of California, San Diego). This year the meeting was held jointly with the new RECOMB Satellite Workshop on Computational Cancer Biology (RECOMB-CCB), and included a joint keynote address by Barbara J. Trask (Fred Hutchinson Cancer Research Center), sponsored by the law firm of Morrison & Foerster, LLP.

RECOMB-CG presentations focus on emerging problems, data, and technologies. This year's invited talks gave particular attention to the evolution of primate genomes and the use of comparative methods for identifying genomic novelties that make us uniquely human. Whole-genome approaches to species tree reconstruction were a dominant theme among the contributed papers. In contrast to the frequently misleading practice of inferring a species tree from a single gene family, comparative genomics research is spawning approaches for deriving phylogenetic signal from entire genomes. Whole-genome methods discussed at RECOMB-CG 2007 included analysis of conserved intron positions and gene order conservation. Another emerging theme was gene duplication. Some papers investigated the role of gene duplication in the evolution of genetic novelties. Other work viewed duplications as a form of uncertainty and proposed methods to address this source of noise in reconstruction of genome rearrangements. A third research thrust discussed at the meeting focussed on novel approaches to inferring ancestral character states, and genome rearrangement distances and phylogenies. Finally, presentations on gene family evolution, as well as the use of comparative methods in inferring regulatory motifs and networks, complemented results on large-scale, spatial genomics.

RECOMB-CG 2007 is indebted to the many individuals and organizations who contributed their support, dedication, and hard work. The Steering Committee supported us in all aspects of the meeting. The success of the meeting depends critically on the efforts of the Program Committee and their sub-reviewers. Their good judgment and constructive criticism engendered an exciting and high-quality scientific program. Paper and poster submission and selection were managed through the EasyChair Web site. We express our appreciation to Andrei Voronkov for providing this system. We are especially grateful to Anita McKee, Jennifer Zimmerman, and Doug Ramsey at UCSD and Annette McLeod at Carnegie Mellon University for administrative support. RECOMB-CG 2007 thanks the law firm Morrison & Foerster, LLP; the National Science Foundation; the University of California's Industry-University Cooperative Research Program; and the University of California, San Diego (UCSD) for financial support and UCSD and the California Institute for Telecommunications and Information Technology (Calit2) for hosting the conference.

Most important, we thank the invited speakers, the scientists who submitted papers and posters, the conference attendees, and the committee members and student volunteers who helped to make this meeting possible. It is the contribution of these individuals that makes RECOMB-CG an exciting scientific event.

September 2007   Glenn Tesler
                                                        Dannie Durand

Conference Organization

Program Committee Chairs

Glenn Tesler (University of California, San Diego, USA)
Dannie Durand (Carnegie Mellon University, USA)

Program Committee

Lars Arvestad (Kungliga Tekniska Högskolan, Sweden)
Vineet Bafna (University of California, San Diego, USA)
Serafim Batzoglou (Stanford University, USA)
Anne Bergeron (Université du Québec à Montréal, Canada)
Mathieu Blanchette (McGill University, Montreal, Canada)
Guillaume Bourque (Genome Institute of Singapore, Singapore)
David Bryant (The University of Auckland, New Zealand)
Jeremy Buhler (Washington University in St. Louis, USA)
Sourav Chatterji (University of California, Berkeley, USA)
Cedric Chauve (Université du Québec à Montréal, Canada)
Avril Coghlan (Wellcome Trust Sanger Institute, UK)
Miklos Csuros (University of Montreal, Canada)
Aaron Darling (University of Queensland, Australia)
Nadia El-Mabrouk (University of Montreal, Canada)
Niklas Eriksen (Chalmers University of Technology, Sweden)
Steffen Heber (North Carolina State University, USA)
Daniel Huson (Eberhard Karls Universität, Tübingen, Germany)
Tao Jiang (University of California, Riverside, USA)
Jens Lagergren (Kungliga Tekniska Högskolan, Sweden)
Aoife McLysaght (Trinity College, University of Dublin, Ireland)
Laxmi Parida (New York University and IBM, USA)
Marie-France Sagot (INRIA, France)
David Sankoff (University of Ottawa, Canada)
Marie Sémon (Université Claude Bernard Lyon 1, France)
Joao Setubal (Virginia Tech, USA)
Jens Stoye (Universität Bielefeld, Germany)
Haixu Tang (Indiana University, USA)
Eric Tannier (INRIA Rhône-Alpes, France)
Tiffani Williams (Texas A & M, USA)
Stacia Wyman (Fred Hutchinson Cancer Research Center, USA)
Liqing Zhang (Virginia Tech, USA)
Louxin Zhang (National University of Singapore, Singapore)
Yves van de Peer (Ghent University, Belgium)

External Reviewers

Laurent Gueguen (Université Claude Bernard Lyon 1, France)
Katharina Jahn (Universität Bielefeld, Germany)
Åsa Pérez-Bercoff (Trinity College, University of Dublin, Ireland)
Roland Wittler (Universität Bielefeld, Germany)
Chunfang Zheng (University of Ottawa, Canada)

Local Organizing Committee

Glenn Tesler (University of California, San Diego, USA)
Mark Chaisson (University of California, San Diego, USA)
Qian Peng (University of California, San Diego, USA)

Steering Committee

Jens Lagergren (Kungliga Tekniska Högskolan, Sweden)
Aoife McLysaght (Trinity College, University of Dublin, Ireland)
David Sankoff (University of Ottawa, Canada)

Sponsors

Morrison & Foerster, LLP (www.mofo.com)
National Science Foundation (www.nsf.gov)
Opportunity Award from the Industry-University Cooperative Research
    Program (www.ucdiscoverygrant.org)
Division of Physical Sciences, University of California, San Diego
    (physicalsciences.ucsd.edu)
Center for Algorithmic and Systems Biology, University of California,
    San Diego (casb.ucsd.edu)
California Institute for Telecommunications and Information
    Technologie (www.calitz.net)

Previous Meetings in This Series

1st RECOMB Satellite Workshop on Comparative Genomics
October 20–24, 2003
Institute for Mathematics and Its Applications (IMA), University of Minnesota,
   Minneapolis, USA
Program Chairs: Jens Lagergren (Stockholm Bioinformatics Centre, KTH,
   Sweden), Bernard Moret (University of New Mexico, USA), and David Sankoff
   (University of Ottawa, Canada)

2nd RECOMB Satellite Workshop on Comparative Genomics
October 16–19, 2004
Bertinoro International Center for Informatics, University of Bologna, Italy
Program Chairs: Jens Lagergren (Stockholm Bioinformatics Centre, KTH,
   Sweden), Aoife McLysaght (Trinity College, Ireland) and David Sankoff
   (University of Ottawa, Canada)

3rd RECOMB Satellite Workshop on Comparative Genomics
September 18–20, 2005
Trinity College, University of Dublin, Ireland
Program Chairs: Aoife McLysaght (Trinity College Dublin, Ireland) and Daniel
   Huson (Eberhard Karls Universität, Tübingen, Germany)

4th RECOMB Satellite Workshop on Comparative Genomics
September 24–26, 2006
University of Montreal, Quebec, Canada
Program Chairs: Nadia El-Mabrouk (University of Montreal, Canada) and
   Guillaume Bourque (Genome Institute of Singapore, Singapore)

Table of Contents

Multi-break Rearrangements: From Circular to Linear Genomes

Max A. Alekseyev

Department of Computer Science and Engineering
University of California at San Diego, USA
maxal@cs.ucsd.edu

Abstract. Multi-break rearrangements break a genome into multiple fragments and further glue them together in a new order. While 2-break rearrangements represent standard reversals, fusions, fissions, and translocations operations; 3-break rearrangements are a natural generalization of transpositions and inverted transpositions. Multi-break rearrangements in circular genomes were studied in depth in [1] and were further applied to the analysis of chromosomal evolution in mammalian genomes [2]. In this paper we extend these results to the more difficult case of linear genomes. In particular, we give lower bounds for the rearrangement distance between linear genomes and use these results to analyze comparative genomic architecture of mammalian genomes.

1 Introduction

Rearrangements are genomic "earthquakes" that change the chromosomal architecture. Each of the standard rearrangement operations (i.e., reversal, translocation, fusion, or fission) can be viewed as making up to 2 breaks in a genome and gluing the resulting fragments in a certain order. More complex rearrangement operations such as transpositions[1] require 3 breaks. Alekseyev and Pevzner [1] introduced a generalized k-break rearrangement operation that makes k breaks in a genomes and glues the resulting fragments in a new order, and studied such operations in depth in the case of *circular genomes* (i.e., genomes consisting of one or more circular chromosomes).

While 2-breaks correspond to the standard rearrangement operations; 3-breaks add transpositions, 3-way fusions, and 3-way fissions to the set of rearrangement operations. Although transpositions are believed to be rare (as compared to reversals and translocations) and 3-way fusions/fissions were never described before in evolutionary context, these complex rearrangements may be involved in chromosome aberrations in irradiated genomes [3,4,5,6]. As shown in [7], switching from transpositions to 3-breaks allows one to attack hard computational problems that otherwise may be intractable. Another application of multi-break rearrangements is the analysis of "FBM vs. RBM" controversy in mammalian evolution [2].

---

[1] Throughout this paper transpositions refer to both transpositions and inverted transpositions.

G. Tesler and D. Durand (Eds.): RECOMB-CG 2007, LNBI 4751, pp. 1–15, 2007.

The *Random Breakage Model (RBM)* [8] postulates that rearrangements happen at "random" genomic positions, resulting in low *breakpoint re-use rate*. RBM was criticized by Pevzner and Tesler, 2003 [9][2] who came up with an alternative *Fragile Breakage Model (FBM)*. The FBM postulates existence of *fragile* genomic regions that are more likely to be broken by rearrangements than the rest of the genome, implying (in contrast to the RBM) high breakpoint re-use rate. A variety of further studies argued for existence of fragile regions in mammalian genomes [12,13,14,15,16,17,18,19,20]. For example, Kikuta et al, 2007 [20] analyzed the links between genome fragility and the need to keep genome intact by regulatory elements and came to the conclusion that "the Nadeau and Taylor hypothesis is not possible for the explanation of synteny in general." However, "RBM vs. FBM" question remains controversial fuelled by recent argument [21] that high breakpoint re-use rate between mammalian genomes maybe caused by more complex rearrangement operations (like transpositions). This concern was addressed in [2] for the case of circular genomes but it remains unclear how to analyze the breakpoint re-use in the more relevant case of linear genomes. While multi-breaks in linear genomes can be defined similarly to circular genomes, the linear case is harder to analyze. In contrast to circular genomes, not every multi-break can be performed over a linear genome: multi-breaks that create circular chromosomes are not allowed. In this paper we extend the results from [1] and [2] to the case of linear genomes. The detailed analysis of biological implications of these results remains beyond the scope of this paper and will be considered elsewhere.

2   Multi-break Rearrangements in Circular Genomes

We will find it convenient to represent a circular chromosome with genes $x_1, \ldots x_n$ as a cycle (Fig. 1) composed of n directed labelled edges (corresponding to genes) and n undirected unlabeled edges (connecting adjacent genes). The directions of the edges correspond to *signs* (strand) of the genes. We label the *tail* and *head* of a directed edge x_i as x_i^t and x_i^h respectively. Vertex x_i^t is called the *obverse* of vertex x_i^h, and vice versa. Vertices in a chromosome connected by an undirected edge are called *adjacent*. We represent a genome as a collection of disjoint cycles (chromosomes) with edges of two *alternating* colors: one color (usually black or gray) reserved for undirected edges and the other ("obverse"[3]) color reserved for directed edges. We do not explicitly show the directions of obverse edges since they are defined by superscripts "t" and "h" (Fig. 1).

Let P be a genome represented as a collection of *alternating cycles* with black and obverse edges (a cycle is alternating if colors of its edges alternate). For any two black edges (u, v) and (x, y) in the genome (graph) P we define a *2-break* rearrangement as replacement of these edges with either a pair of edges (u, x), (v, y),

---

[2] While the rebuttal of RBM caused a controversy [10], recent study [11] revealed an important flaw in arguments supporting RBM [10].

[3] We have chosen rather unusual name "obverse" for the color to be consistent with previous papers on genome rearrangements.

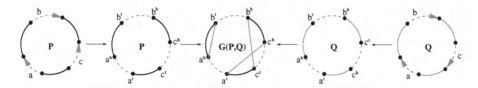

Fig. 1. The breakpoint graph $G(P,Q)$ of unichromosomal genomes $P = +a + b - c$ and $Q = +a + b + c$ represented as a black-obverse cycle and a gray-obverse cycle correspondingly

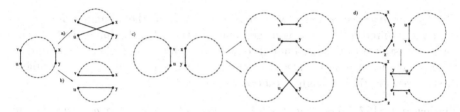

Fig. 2. 2-Break on edges (u,v) and (x,y) corresponding to a) Reversal: the edges belong to the same black-obverse cycle that is rearranged after 2-break; b) Fission: the edges belong to the same black-obverse cycle that is split by 2-break; c) Translocation/fusion: the edges belong to different black-obverse cycles that are joined by 2-break. d) 3-Break on edges (u,v), (x,y) and (z,t) corresponding to transposition of a segment $y \ldots t$ from one chromosome to another. A *transposition* cuts off a segment of one chromosome and inserts it into the same or another chromosome. A transposition of a segment $\pi_i \pi_{i+1} \ldots \pi_j$ of a chromosome $\pi_1 \pi_2 \ldots \pi_i \pi_{i+1} \ldots \pi_j \ldots \pi_k \pi_{k+1} \ldots \pi_m$ into a position k of the same chromosome results a chromosome $\pi_1 \pi_2 \ldots \pi_{i-1} \pi_{j+1} \ldots \pi_k \pi_i \pi_{i+1} \ldots \pi_j \pi_{k+1} \ldots \pi_m$. For chromosomes $\pi = \pi_1 \pi_2 \ldots \pi_i \pi_{i+1} \ldots \pi_j \ldots \pi_m$ and $\sigma = \sigma_1 \sigma_2 \ldots \sigma_n$ a transposition of a segment $\pi_i \pi_{i+1} \ldots \pi_j$ of chromosome π into a position k in the chromosome σ results in chromosomes $\pi_1 \pi_2 \ldots \pi_{i-1} \pi_{j+1} \pi_{j+2} \ldots \pi_m$ and $\sigma_1 \sigma_2 \ldots \sigma_{k-1} \underline{\pi_i \pi_{i+1} \ldots \pi_j} \sigma_k \ldots \sigma_n$.

or a pair of edges (u,y), (v,x). 2-Breaks correspond to standard rearrangement operations of reversals (Fig. 2a), fissions (Fig. 2b), or fusions/translocations[4] (Fig. 2c). 2-Break rearrangements can be generalized as follows. Given k black edges forming a *matching* (i.e., a vertex-disjoint set of edges) on $2k$ vertices, define a *k-break* as replacement of these edges with a set of k black edges forming another matching on the same set of $2k$ vertices. Note that a 2-break is a particular case of a 3-break (as well as of a k-break for $k > 3$), in which case only two edges are replaced and the third one remains the same.

Let P and Q be two signed genomes on the same set of genes $\mathcal{G}$. The *breakpoint graph* $G(P,Q)$ is defined on the set of vertices $V = \{x^t, x^h \mid x \in \mathcal{G}\}$ with edges of three colors: obverse, black, and gray (Fig. 1). Edges of each color form a

---

[4] This definition of elementary rearrangement operations follows the standard definitions of reversals, translocations, fissions, and fusions for the case of circular chromosomes. For circular chromosomes fusions and translocations are not distinguishable.

matching on V: *obverse matching* (pairs of obverse vertices), *black matching* (adjacent vertices in P), and *gray matching* (adjacent vertices in Q). Every pair of matchings forms a collection of alternating cycles in $G(P, Q)$, called *black-gray*, *black-obverse*, and *gray-obverse* cycles respectively. The chromosomes of the genome P (resp. Q) can be read along black-obverse (resp. gray-obverse) cycles. The black-gray cycles in the breakpoint graph play an important role in analyzing rearrangements [22] (see Chapter 10 of [23] for background information on genome rearrangements).

2.1   Multi-break Distance Between Circular Genomes

The *k-break distance* between two genomes is defined as the minimum number of k-breaks required to transform one genome into the other. In difference from the genomic distance (for linear multichromosomal genomes) [24,25,26], computing the 2-break distance for circular multichromosomal genomes is a trivial problem (first solved in [27] in a slightly different context):

Theorem 1 ([27,1,7]). *The 2-break distance between circular genomes P and Q is $d_2(P, Q) = |P| - c(P, Q)$ where $c(P, Q)$ is the number of black-gray cycles in $G(P, Q)$.*

While 2-breaks correspond to standard rearrangements, 3-breaks add transposition-like operations as well as 3-way fissions and fusions to the set of rearrangements (Fig. 2c). In difference from standard rearrangements (modelled as 2-breaks), transpositions introduce 3 breaks in the genome, making them notoriously difficult to analyze. Computing the minimum number of transpositions transforming one genome into another is called *sorting by transpositions*. A number of researchers considered transpositions in conjunction with other rearrangement operations [28,29,30,31,32,33,34]. Despite many studies, the complexity of sorting by transpositions remains unknown [35,36,37,38,39].

Let $c^{odd}(P, Q)$ be the number of black-gray cycles in the breakpoint graph $G(P, Q)$ with an odd number of black edges (*odd cycles*).

Theorem 2 ([1,7]). *The 3-break distance between a black matching P and a gray matching Q is $d_3(P, Q) = \frac{|P| - c^{odd}(P,Q)}{2}$.*

A general formula as well as algorithms for computing the k-break distance can be found in [1].

2.2   Breakpoint Re-Use in Circular Genomes

If each of $d_3(P, Q)$ 3-breaks on a shortest evolutionary path from a circular genome P to a circular genome Q made 3 breaks (*complete* 3-breaks), it would result in the *breakpoint re-use rate* $\frac{3 \cdot d_3(P,Q)}{|P|}$ (i.e., the total number of breaks divided by the number of genes). In reality, some 3-breaks can make 2 breaks (*incomplete* 3-breaks) as 2-breaks are particular cases of 3-breaks, reducing the estimate for the breakpoint re-use rate. Moreover, the minimum breakpoint re-use rate may be achieved on a suboptimal evolutionary path from P to Q.

The rebuttal of RBM raises a question about finding a transformation of one genome into the other by 3-breaks that makes the minimum number of individual breaks. The following theorem shows that there exists a series of $d_3(P, Q)$ 3-breaks that makes the minimum number of breaks while transforming P into Q:

Theorem 3 ([2]). *Any series of m k-breaks transforming a circular genome P into a circular genome Q makes at least $m + d_2(P, Q)$ breaks. Moreover, there exists a series of $d_3(P, Q)$ 3-breaks transforming P into Q that makes $d_3(P, Q) + d_2(P, Q)$ breaks.*

The following theorem shows how the minimum number of breaks in a series of 3-breaks transforming P into Q depends on the number of complete 3-breaks.

Theorem 4 ([2]). *Any series of 3-breaks with t complete 3-breaks, transforming a circular genome P into a circular genome Q, makes at least $d_2(P, Q) + \max\{d_2(P, Q) - t, d_3(P, Q)\}$ breaks. In particular, any such series of 3-breaks with $t \leq d_2(P, Q) - d_3(P, Q)$ complete 3-breaks makes at least $2d_2(P, Q) - t$ breaks.*

In [2] this theorem was applied to the human genome H and the mouse genome M consisting of the 281 synteny blocks from [40] under the assumption that all chromosomes are circular. In the next section we will provide similar estimates for the more relevant case of linear genomes.

The breakpoint graph $G(H, M)$ contains 35 black-gray cycles including 3 odd black-gray cycles, implying that $d_2(H, M) = 281 - 35 = 246$ (Theorem 1) and $d_3(H, M) = 139$ (Theorem 2). Theorem 3 implies that the minimum number of breaks for 2-break sorting H into M is $d_2(H, M) + d_2(H, M) = 246 + 246 = 492$ while for 3-break sorting the minimum number of breaks is $d_3(H, M) + d_2(H, M) = 139 + 246 = 385$.

Figure 3a gives the lower bound for the breakpoint re-use rate as a function of the number of complete 3-breaks in a series of rearrangements transforming H into M. It was argued in [2] that in order to achieve small breakpoint re-use rate between the genomes H and M, the number of complete 3-breaks must be large.

3   Rearrangements in Linear Genomes

A *linear genome* is a collection of linear chromosomes represented as sequences of signed elements (genes). Similarly to circular genomes, we represent each linear chromosome on n genes as a sequence of n directed obverse edges (encoding genes and their direction) and $n-1$ undirected black edges (connecting adjacent genes). So, each linear chromosome is an alternating *path* of obverse and black edges (starting and ending with obverse edges), and a linear genome is a collection of such paths.

Every linear genome P with m chromosomes has $2m$ vertices representing endpoints of the chromosomes. If we introduce an arbitrary perfect matching

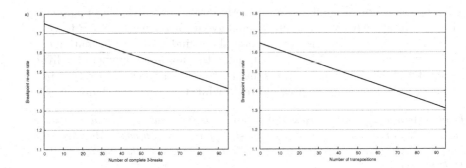

Fig. 3. The lower bound on the breakpoint re-use rate for human and mouse genomes based on 281 synteny blocks from [40]. The lower bound is represented as a function of a) the number of complete 3-breaks in a series of 3-breaks between the circularized human and mouse genomes. (Reproduced from [2]). b) the number of transpositions in a series of rearrangements between the linear human and mouse genomes.

on these $2m$ vertices, consisting of black *closing edges*, the resulting graph will represent some circular genome that contains P as a subgraph. We call the resulting genome a *closure* of P and note that in general it is not uniquely defined. Black edges that belong to P are called *non-closing*.

Throughout this section we assume that P and Q are linear genomes on the same set of genes.

3.1   Rearrangement Distance Between Linear Genomes

Let $d_2^l(P,Q)$ be the genomic distance between the genomes P and Q, i.e., the minimum number of reversals, translocations, fissions, and fusions required to transform P into Q. Also, let $d_3^l(P,Q)$ be the minimum number of reversals, translocations, fissions, and fusions as well as transpositions required to transform P into Q.

Theorem 5. *For any closure P' of a genome P, there exists a closure Q' of a genome Q such that $d_2^l(P,Q) \geq d_2(P',Q')$. Similarly, for any closure P' of a genome P, there exists a closure Q' of a genome Q such that $d_3^l(P,Q) \geq d_3(P',Q')$.*

Proof. Let S' be a closure of a linear genome S. We note that any reversal, translocation, fission, or fusions transforming the genome S into a linear genome T corresponds to a 2-break transforming the closure S' into some closure T' of the genome T (Fig. 4a,b,c,d). Similarly, any transposition transforming the genome S into a linear genome T corresponds to a 3-break transforming the closure S' into some closure T' of the genome T (Fig. 4e).

For the genomes P and Q, consider a series of $d_k^l(P,Q)$ ($k = 2$ or $k = 3$) rearrangements transforming P into Q. This series corresponds to a series of k-breaks transforming P' into some circular genome Q' that is a closure of

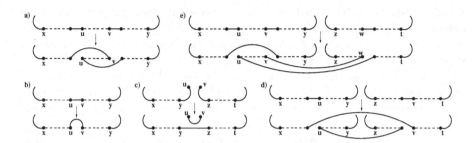

Fig. 4. Rearrangements of linear genomes correspond to k-breaks over closures: a) Reversal of the region (u, v) is a 2-break over non-closing black edges; b) Fission at the black edge (u, v) is the identity multi-break over the edge (u, v), re-claiming this edge as closing; c) Fusion of the chromosomes endpoints y and z is a 2-break replacing closing edges (y, u) and (z, v) with a non-closing edge (y, z) and a closing edge (u, v); d) Translocation exchanging chromosomes parts (u, y) and (v, t) is a 2-break operating over non-closing edges; e) Transposition is a 3-break operating over non-closing edges.

the genome Q. To complete the proof it is sufficient to notice that the distance $d_k(P', Q')$ between the genomes P' and Q' does not exceed $d_k^l(P, Q)$, i.e., $d_k(P', Q') \leq d_k^l(P, Q)$.   □

Theorem 5 immediately implies:

Corollary 6. *For any linear genomes P and Q, $k = 2$ or $k = 3$,*

$$d_k^l(P, Q) \geq \max_{P'} \min_{Q'} d_k(P', Q'),$$
$$d_k^l(P, Q) = d_k^l(Q, P) \geq \max_{Q'} \min_{P'} d_k(P', Q')$$

where P' and Q' vary over all possible closures of the genomes P and Q respectively.

Since the k-break distance $d_k(P', Q')$ $(k = 2$ or $k = 3)$ gives a lower bound for the linear distance $d_k^l(P, Q)$, our goal is to make this bound as tight as possible by choosing appropriate closures P' and Q'. We start with defining the breakpoint graph of linear genomes and a number of its characteristics that we will find useful.

Let P' and Q' be closures of linear genomes P and Q. The breakpoint graph $G(P, Q)$ is defined as a result of removal of all closing edges from the breakpoint graph $G(P', Q')$ (of circular genomes P' and Q'). It is easy to see that $G(P, Q)$ is well-defined by the genomes P and Q and does not depend on a particular choice of closures P' and Q'. Every cycle in $G(P', Q')$ with m closing edges will be split into m paths in $G(P, Q)$. Therefore, the black-gray connected components of $G(P, Q)$ are formed by $c(P, Q)$ black-gray cycles and a number of black-gray paths. We distinguish between black-gray paths with both terminal edges of black color (*bb-paths*), with both terminal edges of gray color (*gg-paths*), and with terminal edges of different colors (*bg-paths*), including isolated vertices viewed as bg-paths with zero black and zero gray edges. We denote the number of such paths by $l_{bb}(P, Q)$, $l_{gg}(P, Q)$, and $l_{bg}(P, Q)$ respectively (note that the number $l_{bg}(P, Q)$ is always even). The total number of black-gray connected components in $G(P, Q)$ is

$$cc(P, Q) = c(P, Q) + l_{bb}(P, Q) + l_{gg}(P, Q) + l_{bg}(P, Q).$$

We also distinguish between black-gray connected components with odd/even number of black/gray edges and call them *b-odd*, *b-even*, *g-odd*, *g-even* respectively. To refer to the number of such components we will use these aliases as superscripts. Similarly to cycles, bg-paths have the same number of black and gray edges, so we call bg-paths simply *odd* and *even*, depending on the oddness of the number of black edges. Relatedly, $l_{bg}^{odd}(P,Q)$ and $l_{bg}^{even}(P,Q)$ will stand for the number of odd and even bg-paths. We rely on the following identities:

$$\forall j \in \{bb, bg, gg\},$$
$$l_j(P,Q) = l_j^{b-odd}(P,Q) + l_j^{b-even}(P,Q),\ l_j(P,Q) = l_j^{g-odd}(P,Q) + l_j^{g-even}(P,Q);$$
$$\forall j \in \{bb, gg\},$$
$$l_j^{b-odd}(P,Q) = l_j^{g-even}(P,Q), \qquad l_j^{b-even}(P,Q) = l_j^{g-odd}(P,Q).$$

These identities allow to compute all the characteristics defined above as soon as $c(P,Q)$, $c^{odd}(P,Q)$, $l_{bb}(P,Q)$, $l_{bb}^{b-odd}(P,Q)$, $l_{gg}(P,Q)$, $l_{gg}^{b-odd}(P,Q)$, $l_{bg}(P,Q)$, $l_{bg}^{odd}(P,Q)$ are given.

Similarly to the breakpoints graphs for circular and linear genomes, we can define the breakpoint graph and associated characteristics in the case when one genome is circular while the other is linear (in such a graph all paths are either bb-paths or gg-paths).

Lemma 7. *For a circular genome P' and a linear genome Q,*

$$\min_{Q'} d_2(P',Q') = |P'| - cc(P',Q) \quad and \quad \min_{Q'} d_3(P',Q') = \frac{|P'| - cc^{b-odd}(P',Q)}{2}$$

where Q' varies over all possible closures of the genome Q.

Proof. Theorem 1 implies that $\min_{Q'} d_2(P',Q') = |P'| - \max_{Q'} c(P',Q')$. To maximize $c(P',Q')$, the closure Q' needs to be chosen in such a way that it closes each path in the breakpoint graph $G(P',Q)$ into a separate black-gray cycle. Therefore, $\max_{Q'} c(P',Q') = cc(P',Q)$.

Similarly, Theorem 2 implies that

$$\min_{Q'} d_3(P',Q') = \frac{|P'| - \max_{Q'} c^{odd}(P',Q')}{2}.$$

To maximize $c^{odd}(P',Q')$, the closure Q' needs to be chosen in such a way that it closes each b-odd path in the breakpoint graph $G(P',Q)$ into a separate black-gray cycle. Therefore, $\max_{Q'} c(P',Q') = cc^{b-odd}(P',Q)$.   □

Theorem 8. *For linear genomes P and Q, $\max_{P'} \min_{Q'} d_2(P',Q') = B_2(P,Q)$ where*

$$B_2(P,Q) = |P| - c(P,Q) - \max\{1, \frac{l_{bg}(P,Q)}{2}\} - l_{bb}(P,Q),$$

implying that $d_2^l(P,Q) \geq \max\{B_2(P,Q), B_2(Q,P)\}$.

Proof. By Lemma 7 we have $\max_{P'} \min_{Q'} d_2(P', Q') = |P| - \min_{P'} cc(P', Q)$. In order to minimize $cc(P', Q)$, the closure P' needs to be chosen in such a way that it minimizes the number of black-gray connected components in $G(P, Q)$. This can be done as follows. If $l_{bg}(P, Q) = 0$, then we will connect (using closing black edges) all the gg-paths into a single cycle. If $l_{bg}(P, Q) > 0$, we will first connect a pair of bg-paths and all the gg-paths into a single bb-path, and then form pairs of the remaining bg-paths and connect bg-paths in each pair into a bb-path. As a result, $\min_{P'} cc(P', Q) = c(P, Q) + \max\{1, \frac{l_{bg}(P,Q)}{2}\} + l_{bb}(P, Q)$. Therefore, $\max_{P'} \min_{Q'} d_2(P', Q') = B_2(P, Q)$ and by Corollary 6, $d_2^l(P, Q) \geq B_2(P, Q)$. Moreover, since $d_2^l(P, Q) = d_2^l(Q, P) \geq B_2(Q, P)$, we have $d_2^l(P, Q) \geq \max\{B_2(P, Q), B_2(Q, P)\}$.   □

Lemma 9. *For linear genomes P and Q, $\min_{P'} cc^{b-odd}(P', Q) = L_3(P, Q)$ where*

$$L_3(P, Q) = c^{odd}(P, Q) + l_{bb}^{b-odd}(P, Q) + \delta(P, Q)$$
$$+ \max\left\{0, \frac{|l_{bg}^{odd}(P,Q) - l_{bg}^{even}(P,Q)|}{2} - l_{gg}^{b-even}(P, Q)\right\}$$

and $\delta(P, Q) = \max\left\{0, l_{gg}^{b-even}(P, Q) - \frac{|l_{bg}^{odd}(P,Q) - l_{bg}^{even}(P,Q)|}{2}\right\}$ mod 2.

Proof. Note that in any closure of P, the closing (black) edges connect gg-paths and bg-paths from $G(P, Q)$ into $m_1 = \frac{l_{bg}(P,Q)}{2}$ bb-paths and a number of cycles. Note that if $l_{bg}(P, Q) = 0$ then connecting all gg-paths into a single cycle (which will be odd iff $l_{gg}^{b-even}(P, Q)$ is odd) gives an optimal closure P'' (i.e., for which $\min_{P'} cc^{b-odd}(P', Q) = cc^{b-odd}(P'', Q)$). It is easy to check that in this case $cc^{b-odd}(P'', Q) = L_3(P, Q)$. For the rest of the proof we assume that $l_{bg}(P, Q) > 0$.

We will show that there exists an optimal closure where the closing edges do not connect any gg-paths into a cycle. Such an optimal closure can be obtained from an arbitrary optimal closure P'' as explained below. Since $l_{bg}(P, Q) > 0$, the closing edges in $G(P'', Q)$ create at least one bb-path formed by two bg-paths at the ends and possibly gg-paths in the middle. Let us re-connect (modifying the set of closing edges) all the gg-paths from $G(P, Q)$, that are connected into cycles in $G(P'', Q)$, in the middle of this bb-path. Note that such modification of the closure may change the b-oddness of the affected bb-path but only if at least one of the destroyed cycles was odd. In any case the number of b-odd connected components is not increased. Therefore, the modified closure is optimal and satisfies the required property by construction. Without loss of generality we will assume that the closing edges create no cycles.

Bringing black closing edges into $G(P, Q)$ can be viewed as a two-step process: first, connecting gg-paths into longer gg-paths; and second, connecting pairs of bg-paths and maybe single gg-paths into bb-paths. Our goal is to minimize the number of b-odd bb-paths or, equivalently, to maximize the number of b-even bb-paths.

Consider an outcome of the first step. It is clear that connection of two b-odd gg-paths or two b-even gg-paths results in a b-odd gg-path, while connection of b-odd and b-even gg-paths results in b-even gg-path. As we will see b-even gg-paths are more preferable than b-odd gg-paths. After the first step we can have up to $m_2 = l_{gg}^{b-even}(P,Q)$ b-even gg-paths.

Now, consider the second step. Connection of an odd bg-path and an even bg-path with an optional b-odd gg-path in between create a b-even bb-path. At the same time connection of a pair of odd bg-paths or a pair of even bg-paths requires a b-even gg-path in between in order to produce a b-even bb-path. All other combinations of bg-paths and gg-paths result in b-odd bb-paths.

We can create $m_3 = \min\{l_{bg}^{odd}(P,Q), l_{bg}^{even}(P,Q)\}$ b-even bb-paths without any use of gg-paths, and up to $m_4 = \frac{|l_{bg}^{odd}(P,Q) - l_{bg}^{even}(P,Q)|}{2}$ b-even bb-paths (note that $m_3 + m_4 = m_1$), each of which requires a b-even gg-path in the middle. Hence, we can create $m_5 = m_3 + \min\{m_4, m_2\} = \min\{m_1, m_2 + m_3\}$ b-even bb-paths. The other $m_6 = m_1 - m_5 = \max\{0, m_4 - m_2\}$ bb-paths (formed by pairs of bg-paths of the same oddness) will be b-odd. So far we have used $\min\{m_4, m_2\}$ b-even gg-paths. The other gg-paths (if any) can be connected (at the first step) into a single gg-path that is b-odd iff $m_2 - \min\{m_4, m_2\} = \max\{m_2 - m_4, 0\}$ is odd (i.e., $\delta(P,Q) = 1$). The b-odd gg-path can be easily incorporated into any of created bb-paths without changing its b-oddness. The b-even gg-path we have to incorporate into some of created b-even bb-paths and turn it into a b-odd bb-path. Hence, for an optimal closure P', there are $l_{bb}^{b-odd}(P,Q) + m_6 + \delta(P,Q)$ b-odd bb-paths and $c^{odd}(P,Q)$ odd cycles in $G(P',Q)$, implying that $cc^{b-odd}(P',Q) = c^{odd}(P,Q) + l_{bb}^{b-odd}(P,Q) + m_6 + \delta(P,Q)$. $\quad\square$

Theorem 10. *For linear genomes P and Q, $d_3^l(P,Q) \geq \max\{B_3(P,Q), B_3(Q,P)\}$ where $B_3(P,Q) = \frac{|P| - L_3(P,Q)}{2}$.*

Proof. Since $d_3^l(P,Q) = d_3^l(Q,P)$ it is sufficient to show that $d_3^l(P,Q) \geq B_3(P,Q)$. Corollary 6 and Lemma 7 imply

$$d_3^l(P,Q) \geq \max_{P'} \min_{Q'} d_3(P',Q') = \frac{|P| - \min_{P'} cc^{b-odd}(P',Q)}{2}.$$

Now, application of Lemma 9 completes the proof. $\quad\square$

3.2   Breakpoint Re-Use in Linear Genomes

Similarly to the case of circular genomes, we are interested in estimating the total number breaks required to transform a linear genome P into a linear genome Q with reversals, fusions, fissions, translocations, and transpositions. According to Theorem 5, any series of such rearrangements corresponds to a series of 3-breaks transforming a closure P' of the genome P into some closure Q' of the genome Q. Let $b^c(P,Q)$ be the minimum number of breaks made in such a series of 3-breaks (over all possible closures P' and Q'). Theorems 5 and 3 imply:

Corollary 11. *For linear genomes P and Q,*

$$b^c(P,Q) \geq \max_{P'} \min_{Q'} d_3(P',Q') + d_2(P',Q'),$$
$$b^c(P,Q) = b^c(Q,P) \geq \max_{Q'} \min_{P'} d_3(P',Q') + d_2(P',Q')$$

where P' and Q' vary over all possible closures of the genomes P and Q respectively.

To find out the exact value of $\max_{P'} \min_{Q'} d_3(P', Q') + d_2(P', Q')$ we need the following lemma:

Lemma 12. *For a circular genome P' and a linear genome Q,*

$$\min_{Q'} d_2(P', Q') + d_3(P', Q') = \frac{3}{2}|P'| - \frac{3cc^{b-odd}(P', Q) + 2cc^{b-even}(P', Q)}{2}.$$

Proof. Theorems 1 and 2 imply that

$$\min_{Q'} d_3(P', Q') + d_2(P', Q') = \frac{3}{2}|P'| - \frac{\max_{Q'} 3c^{odd}(P', Q') + 2c^{even}(P', Q')}{2}.$$

To maximize $3c^{odd}(P', Q') + 2c^{even}(P', Q')$, a closure Q' has to be chosen in such a way that it closes each path in the breakpoint graph $G(P', Q)$, into a separate black-gray cycle. Indeed, having $m > 1$ paths connected into a single cycle is always worse than connecting each of these paths into a separate cycle as $3 < 2m$. Therefore, for an optimal closure Q', we have $c^{odd}(P', Q') = cc^{b-odd}(P', Q)$ and $c^{even}(P', Q') = cc^{b-even}(P', Q)$. □

Theorem 13. *For linear genomes P and Q, $\max_{P'} \min_{Q'} d_3(P', Q') + d_2(P', Q') = B_{23}(P, Q)$ where*

$$B_{23}(P, Q) = \frac{3}{2}|P| - c(P, Q) - l_{bb}(P, Q) - \frac{l_{bg}(P, Q) + L_3(P, Q)}{2},$$

implying that $b^c(P, Q) \geq \max\{B_{23}(P, Q), B_{23}(Q, P)\}$.

Proof. By Lemma 12 we have

$$\max_{P'} \min_{Q'} d_3(P', Q') + d_2(P', Q') = \frac{3}{2}|P| - \frac{\min_{P'} 3cc^{b-odd}(P', Q) + 2cc^{b-even}(P', Q)}{2}.$$

Note that if $l_{bg}(P, Q) = 0$ then connecting all gg-paths into a single cycle (which will be odd iff $l_{gg}^{b-even}(P, Q)$ is odd) gives an optimal closure P'. It is easy to check that in this case $\max_{P'} \min_{Q'} d_3(P', Q') + d_2(P', Q') = B_{23}(P, Q)$. For the rest of the proof we assume that $l_{bg}(P, Q) > 0$.

Note that in any closure of P, the closing (black) edges connect gg-paths and bg-paths from $G(P, Q)$ into bb-paths and cycles. We will show that in an optimal closure the closing edges do not connect any gg-paths into a cycle. Indeed, since $l_{bg}(P, Q) > 0$, the closing edges create at least one bb-path formed by two bg-paths at the ends and possibly gg-paths in the middle. It is easy to see that it is always better to include more gg-paths in the middle of this bb-path (maybe letting the objective function increase by one) rather than to create a separate cycle out of these gg-paths (in which case the objective function would increase by at least 2). Therefore, closing edges in an optimal closure P' connect gg-paths and bg-paths from $G(P,Q)$ into $\frac{l_{bg}(P,Q)}{2}$ bb-paths in $G(P',Q)$. As the total number of new bb-paths is fixed, the problem of minimizing $3cc^{b-odd}(P',Q) + 2cc^{b-even}(P',Q)$ is equivalent to minimizing $cc^{b-odd}(P',Q)$. For an optimal closure P', Lemma 9 gives $cc^{b-odd}(P',Q) = L_3(P,Q)$, implying that

$$3cc^{b-odd}(P',Q) + 2cc^{b-even}(P',Q) = 2cc(P',Q) + cc^{b-odd}(P',Q)$$
$$= 2(c(P,Q) + l_{bb}(P,Q) + \tfrac{l_{bg}(P,Q)}{2}) + L_3(P,Q)$$

and thus $\max_{P'} \min_{Q'} d_3(P',Q') + d_2(P',Q') = B_{23}(P,Q)$.

By Corollary 11 we have $b^c(P,Q) \geq B_{23}(P,Q)$ and $b^c(P,Q) \geq B_{23}(Q,P)$, implying that $b^c(P,Q) \geq \max\{B_{23}(P,Q), B_{23}(Q,P)\}$. □

We will now prove the following analog of Theorem 4:

Theorem 14. *Any series of rearrangements with t transpositions, transforming a linear genome P into a linear genome Q, makes at least $\max\{2B_2(P,Q) - t, B_{23}(P,Q)\} - \mathrm{chr}(P) + \mathrm{chr}(Q)$ breaks, where $\mathrm{chr}(\cdot)$ denotes the number of chromosomes. In particular, any such series of rearrangements with $t \leq 2B_2(P,Q) - B_{23}(P,Q)$ transpositions makes at least $2B_2(P,Q) - \mathrm{chr}(P) + \mathrm{chr}(Q) - t$ breaks.*

Proof. By Theorem 5, any series of rearrangements transforming the genome P into the genome Q corresponds to a series of 3-breaks transforming a closure P' of P into some closure Q' of Q. We note that every rearrangement makes the same number of breaks as the corresponding 3-break in the closures;[5] except for fusions that make smaller number of breaks than the corresponding 2-breaks in the closures (Fig. 4b), and for fissions that make breaks in linear genomes but correspond to identity multi-breaks (making no breaks) in their closures (Fig. 4c).

Let u, v, t be respectively the number of fusions, fissions, and transpositions in a series of m rearrangements transforming the genome P into the genome Q and making b breaks in total. Then there is a series of 3-breaks, transforming a closure P' into a closure Q', that makes $b + u - v$ breaks in total. Since every fusion decreases the number of chromosomes by one, while every fission increases the number of chromosomes by one, $u - v = \mathrm{chr}(P) - \mathrm{chr}(Q)$. By Theorem 4,

$$b + u - v = b + \mathrm{chr}(P) - \mathrm{chr}(Q) \geq d_2(P',Q') + \max\{d_2(P',Q') - t, d_3(P',Q')\},$$

implying that $b \geq \max\{2d_2(P',Q') - t, d_2(P',Q') + d_3(P',Q')\} + \mathrm{chr}(Q) - \mathrm{chr}(P)$. Taking $\max_{P'} \min_{Q'}$ of the right hand side of this inequality, we have $b \geq \max\{2B_2(P,Q) - t, B_{23}(P,Q)\} + \mathrm{chr}(Q) - \mathrm{chr}(P)$. □

---

[5] We assume that a transposition always makes 3 breaks even if it transposes a part of chromosome starting with one of its ends, a translocation always makes 2 breaks even if it exchanges an entire chromosome with a part of another chromosome, and a reversal always makes 2 breaks even if it involves an end of a chromosome. The biological rationale for this assumption is that chromosomes are flanked by telomeres that while remaining "invisible" in genomic sequences, can account for breakpoint re-use in the same way as any other genomic position.

Using 281 synteny blocks between the linear human genome H and mouse genome M from [40], we estimate the breakpoint re-use rate across these (linear) genomes. The breakpoint graph $G(H, M)$ have the following parameters: $(c, l_{bb}, l_{gg}, l_{bg}) = (28, 12, 15, 16)$, $(c^{odd}, l_{bb}^{b-odd}, l_{gg}^{b-odd}, l_{bg}^{odd}) = (2, 5, 4, 3)$, $\mathrm{chr}(H) = 23$, $\mathrm{chr}(M) = 20$, $B_2(H, M) = 233$, $B_2(M, H) = 230$, $B_3(H, M) = 137$, $B_3(M, H) = 134$, $B_{23}(H, M) = 370$, and $B_{23}(M, H) = 364$. Theorems 8 and 10 imply that $d_2^l(H, M) \geq 233$ and $d_3^l(H, M) \geq 137$.

Theorem 14 gives the lower bound for the breakpoint re-use rate (as a function of the number of transpositions) between the genomes H and M, shown in Fig. 3b. This illustrates that a very large number of transpositions would be necessary to bring the breakpoint re-use rate below the 1.25 rate expected for RBM (see [9,2]). Therefore, Sankoff's argument that high breakpoint re-use rate reported for human-mouse genomic architectures is an artifact caused by not accounting for complex rearrangements [21] may only hold if one assumes that transpositions are dominant rearrangement operations that are more frequent than reversals, translocations, fissions, and fusions. While detailed analysis of such an extreme rearrangement scenario remains beyond the scope of this paper we remark that currently there is no biological evidence to support this scenario.

References

1. Alekseyev, M.A., Pevzner, P.A.: Multi-Break Rearrangements and Chromosomal Evolution. Theoretical Computer Science (to appear, 2007)
2. Alekseyev, M.A., Pevzner, P.A.: Are There Rearrangement Hotspots in the Human Genome? PLos Computational Biology (to appear, 2007)
3. Sachs, R.K., Levy, D., Hahnfeldt, P., Hlatky, L.: Quantitative analysis of radiation-induced chromosome aberrations. Cytogenetic and Genome Research 104, 142–148 (2004)
4. Levy, D., Vazquez, M., Cornforth, M., Loucas, B., Sachs, R.K., Arsuaga, J.: Comparing DNA damage-processing pathways by computer analysis of chromosome painting data. J. Comput. Biol. 11, 626–641 (2004)
5. Vazquez, M., et al.: Computer analysis of mFISH chromosome aberration data uncovers an excess of very complicated metaphases. Int. J. Radiat. Biol. 78(12), 1103–1115 (2002)
6. Sachs, R.K., Arsuaga, J., Vazquez, M., Hlatky, L., Hahnfeldt, P.: Using graph theory to describe and model chromosome aberrations. Radiat Research 158, 556–567 (2002)
7. Alekseyev, M.A., Pevzner, P.A.: Whole Genome Duplications, Multi-Break Rearrangements, and Genome Halving Theorem. Proceedings of the 18th Annual ACM-SIAM Symposium on Discrete Algorithms (SODA) , 665–679 (2007)
8. Nadeau, J.H., Taylor, B.A.: Lengths of Chromosomal Segments Conserved since Divergence of Man and Mouse. Proceedings of the National Academy of Sciences 81(3), 814–818 (1984)
9. Pevzner, P.A., Tesler, G.: Human and mouse genomic sequences reveal extensive breakpoint reuse in mammalian evolution. Proceedings of the National Academy of Sciences 100, 7672–7677 (2003)
10. Sankoff, D., Trinh, P.: Chromosomal breakpoint re-use in the inference of genome sequence rearrangement. In: Proceedings of the Eighth Annual International Conference on Computational Molecular Biology (RECOMB), pp. 30–35 (2004)

11. Peng, Q., Pevzner, P.A., Tesler, G.: The Fragile Breakage versus Random Breakage Models of Chromosome Evolution. PLoS Comput. Biol. 2, e14 (2006)
12. Murphy, W.J., Larkin, D.M., van der Wind, A.E., Bourque, G., Tesler, G., Auvil, L., Beever, J.E., Chowdhary, B.P., Galibert, F., Gatzke, L., Hitte, C., Meyers, C.N., Milan, D., Ostrander, E.A., Pape, G., Parker, H.G., Raudsepp, T., Rogatcheva, M.B., Schook, L.B., Skow, L.C., Welge, M., Womack, J.E., OBrien, S.J., Pevzner, P.A., Lewin, H.A.: Dynamics of Mammalian Chromosome Evolution Inferred from Multispecies Comparative Map. Science 309(5734), 613–617 (2005)
13. van der Wind, A.E., Kata, S.R., Band, M.R., Rebeiz, M., Larkin, D.M., Everts, R.E., Green, C.A., Liu, L., Natarajan, S., Goldammer, T., Lee, J.H., McKay, S., Womack, J.E., Lewin, H.A.: A 1463 Gene Cattle-Human Comparative Map With Anchor Points Defined by Human Genome Sequence Coordinates. Genome Research 14(7), 1424–1437 (2004)
14. Bailey, J., Baertsch, R., Kent, W., Haussler, D., Eichler, E.: Hotspots of mammalian chromosomal evolution. Genome Biology 5(4), R23 (2004)
15. Zhao, S., Shetty, J., Hou, L., Delcher, A., Zhu, B., Osoegawa, K., de Jong, P., Nierman, W.C., Strausberg, R.L., Fraser, C.M.: Human, Mouse, and Rat Genome Large-Scale Rearrangements: Stability Versus Speciation. Genome Research 14, 1851–1860 (2004)
16. Webber, C., Ponting, C.P.: Hotspots of mutation and breakage in dog and human chromosomes. Genome Research 15(12), 1787–1797 (2005)
17. Hinsch, H., Hannenhalli, S.: Recurring genomic breaks in independent lineages support genomic fragility. BMC Evolutionary Biology 6, 90 (2006)
18. Ruiz-Herrera, A., Castresana, J., Robinson, T.J.: Is mammalian chromosomal evolution driven by regions of genome fragility? Genome Biology 7, R115 (2006)
19. Mehan, M.R., Almonte, M., Slaten, E., Freimer, N.B., Rao, P.N., Ophoff, R.A.: Analysis of segmental duplications reveals a distinct pattern of continuation-of-synteny between human and mouse genomes. Human Genetics 121(1), 93–100 (2007)
20. Kikuta, H., Laplante, M., Navratilova, P., Komisarczuk, A.Z, Engstrom, P.G., Fredman, D., Akalin, A., Caccamo, M., Sealy, I., Howe, K., Ghislain, J., Pezeron, G., Mourrain, P., Ellingsen, S., Oates, A.C., Thisse, C., Thisse, B., Foucher, I., Adolf, B., Geling, A., Lenhard, B., Becker, T.S.: Genomic regulatory blocks encompass multiple neighboring genes and maintain conserved synteny in vertebrates. Genome Research 17(5), 545–555 (2007)
21. Sankoff, D.: The signal in the genome. PLoS Computational Biology 2(4), 320–321 (2006)
22. Bafna, V., Pevzner, P.A.: Genome rearrangement and sorting by reversals. SIAM Journal on Computing 25, 272–289 (1996)
23. Pevzner, P.A.: Computational Molecular Biology: An Algorithmic Approach. MIT Press, Cambridge (2000)
24. Hannenhalli, S., Pevzner, P.: Transforming men into mouse (polynomial algorithm for genomic distance problem). In: Proceedings of the 36th Annual Symposium on Foundations of Computer Science, pp. 581–592 (1995)
25. Tesler, G.: Efficient algorithms for multichromosomal genome rearrangements. J. Comput. Syst. Sci. 65, 587–609 (2002)
26. Ozery-Flato, M., Shamir, R.: Two Notes on Genome Rearrangement. Journal of Bioinformatics and Computational Biology 1, 71–94 (2003)
27. Yancopoulos, S., Attie, O., Friedberg, R.: Efficient sorting of genomic permutations by translocation, inversion and block interchange. Bioinformatics 21, 3340–3346 (2005)

28. Bader, M., Ohlebusch, E.: Sorting by weighted reversals, transpositions, and inverted transpositions. In: Apostolico, A., Guerra, C., Istrail, S., Pevzner, P., Waterman, M. (eds.) RECOMB 2006. LNCS (LNBI), vol. 3909, pp. 563–577. Springer, Heidelberg (2006)

29. Gu, Q.P., Peng, S., Sudborough, H.: A 2-approximation algorithm for genome rearrangements by reversals and transpositions. Theoret. Comput. Sci. 210, 327–339 (1999)

30. Hartman, T., Sharan, R.: A 1.5-approximation algorithm for sorting by transpositions and transreversals. In: Jonassen, I., Kim, J. (eds.) WABI 2004. LNCS (LNBI), vol. 3240, pp. 50–61. Springer, Heidelberg (2004)

31. Lin, G.H., Xue, G.: Signed genome rearrangements by reversals and transpositions: models and approximations. Theoret. Comput. Sci. 259, 513–531 (2001)

32. Lin, Y.C., Lu, C.L., Chang, H.-Y., Tang, C.Y.: An Efficient Algorithm for Sorting by Block-Interchanges and Its Application to the Evolution of Vibrio Species. J. Comput. Biol. 12, 102–112 (2005)

33. Radcliffe, A.J., Scott, A.D., Wilmer, E.L.: Reversals and Transpositions Over Finite Alphabets. SIAM J. Discrete Math. 19, 224–244 (2005)

34. Walter, M.E., Dias, Z., Meidanis, J.: Reversal and transposition distance of linear chromosomes. In: String Processing and Information Retrieval: A South American Symposium (SPIRE), pp. 96–102 (1998)

35. Bafna, V., Pevzner, P.A.: Sorting permutations by transpositions. SIAM J. Discrete Math. 11, 224–240 (1998)

36. Christie, D.A.: Genome Rearrangement Problems. PhD thesis, University of Glasgow (1999)

37. Walter, M.E., Reginaldo, L., Curado, A.F., Oliveira, A.G.: Working on the Problem of Sorting by Transpositions on Genome Rearrangements. In: Baeza-Yates, R.A., Chávez, E., Crochemore, M. (eds.) CPM 2003. LNCS, vol. 2676, pp. 372–383. Springer, Heidelberg (2003)

38. Hartman, T.: A simpler 1.5-approximation algorithm for sorting by transpositions. In: Baeza-Yates, R.A., Chávez, E., Crochemore, M. (eds.) CPM 2003. LNCS, vol. 2676, pp. 156–169. Springer, Heidelberg (2003)

39. Elias, I., Hartman, T.: A 1.375-Approximation Algorithm for Sorting by Transpositions. In: Casadio, R., Myers, G. (eds.) WABI 2005. LNCS (LNBI), vol. 3692, pp. 204–214. Springer, Heidelberg (2005)

40. Pevzner, P., Tesler, G.: Genome Rearrangements in Mammalian Evolution: Lessons from Human and Mouse Genomes. Genome Research 13(1), 37–45 (2003)

A Pseudo-boolean Programming Approach for Computing the Breakpoint Distance Between Two Genomes with Duplicate Genes

Sébastien Angibaud[1], Guillaume Fertin[1], Irena Rusu[1], Annelyse Thévenin[2], and Stéphane Vialette[2]

[1] Laboratoire d'Informatique de Nantes-Atlantique (LINA), FRE CNRS 2729
Université de Nantes, 2 rue de la Houssinière, 44322 Nantes Cedex 3 - France
{angibaud,fertin,rusu}@lina.univ-nantes.fr
[2] Laboratoire de Recherche en Informatique (LRI), UMR CNRS 8623
Faculté des Sciences d'Orsay - Université Paris-Sud, 91405 Orsay - France
{thevenin,vialette}@lri.fr

Abstract. Comparing genomes of different species has become a crucial problem in comparative genomics. Recent research have resulted in different genomic distance definitions: number of breakpoints, number of common intervals, number of conserved intervals, Maximum Adjacency Disruption number (MAD), etc. Classical methods (usually based on permutations of gene order) for computing genomic distances between whole genomes are however seriously compromised for genomes where several copies of the same gene may be scattered across the genome. Most approaches to overcoming this difficulty are based on the *exemplar* method (keep exactly one copy in each genome of each duplicated gene) and the *maximum matching* method (keep as many copies as possible in each genome of each duplicated gene). Unfortunately, it turns out that, in presence of duplications, most problems are **NP**–hard, and hence several heuristics have been recently proposed.

Extending research initiated in [2], we propose in this paper a novel generic pseudo-boolean approach for computing the exact breakpoint distance between two genomes in presence of duplications for both the *exemplar* and *maximum matching* methods. We illustrate the application of this methodology on a well-known public benchmark dataset of γ-Proteobacteria.

Keywords: genome rearrangement, duplication, breakpoint distance, heuristic, pseudo-boolean programming.

1 Introduction

The order of genes in the genomes of species can change during evolution and can provide information about their phylogenetic relationship. Two main approaches are possible. The first one consists in using different types of rearrangement operations and to find possible rearrangement scenarios using these operations (one

G. Tesler and D. Durand (Eds.): RECOMB-CG 2007, LNBI 4751, pp. 16–29, 2007.
© Springer-Verlag Berlin Heidelberg 2007

of the most common rearrangement operations is reversals, which reverse the order of a subset of neighboring genes) [11]. The second one consists in computing a (dis-)similarity measure based on the gene order and most common rearrangement operations [15,8,4,1]. We focus in this paper on the latter approach.

Several similarity (or dissimilarity) measures between two whole genomes have been recently proposed, such as the number of breakpoints [15,8,4], the number of reversals [8,11], the number of conserved intervals [6], the number of common intervals [7], the Maximum Adjacency Disruption Number (MAD) [16], *etc.* However, in the presence of duplications and for each of the above measures, one has first to disambiguate the data by inferring orthologs, *i.e.*, a non-ambiguous mapping between the genes of the two genomes. Up to now, two extremal approaches have been considered : the *exemplar* model and the *maximum matching* model. In the *exemplar* model [15], for all gene families, all but one occurrence in each genome is deleted. In the *maximum matching* model [4,10], the goal is to map as many genes as possible. These two models can be considered as the extremal cases of the same generic homolog assignment approach.

Unfortunately, it has been shown that, for each of the above mentioned measures, whatever the considered model (*exemplar* or *maximum matching*), the problem becomes **NP**–complete as soon as duplicates are present in genomes [8,4,6,10] ; a few inapproximability results are known for some special cases. Therefore, several heuristic methods have been recently devised to obtain (hopefully) good solutions in a reasonable amount of time [5,7]. However, while it is relatively easy to compare heuristics between them, until now very little is known about the absolute accuracy of these heuristics. Therefore, there is a great need for algorithmic approaches that compute exact solutions for these genomic distances.

Extending research initiated in [2], we propose in this paper a novel generic pseudo-boolean approach for computing the exact breakpoint distance between two genomes in presence of duplications for both the *exemplar* and *maximum matching* methods. Furthermore, we show strong evidence that a fast and simple heuristic based on iteratively finding longest common subsequences provides very good results on our dataset of γ-Proteobacteria.

This paper is organized as follows. In Section 2, we present some preliminaries and definitions. We focus in Section 3 on the problem of finding the minimum number of breakpoints under the two models and we give a pseudo-boolean program together with some reduction rules. Section 4 is devoted to experimental results on a dataset of γ-Proteobacteria.

2   Preliminaries

From an algorithmic perspective, a *unichromosomal genome* is a signed sequence over a finite alphabet, referred hereafter as the alphabet of *gene families*. Each element of the sequence is called a *gene*. DNA has two strands, and genes on a genome have an orientation that reflects the strand of the genes. We represent the order and directions of the genes on each genome as a sequence of signed elements, *i.e.*, elements with signs "+" and "−". Let G_0 and G_1 be two genomes. For each $x \in \{0, 1\}$, we denote the label at position i in G_x by $G_x[i]$, $1 \leq i \leq n_x$, and we write n_x for the number of genes in genome G_x and $\text{occ}_x(\mathbf{g}, i, j)$ for the number of genes $\mathbf{g}$ (and $-\mathbf{g}$) in G_x between positions i and j, $1 \leq i < j \leq n_x$. To simplify notations, we abbreviate $\text{occ}_x(\mathbf{g}, 1, n_x)$ to $\text{occ}_x(\mathbf{g})$.

In order to deal with the inherent ambiguity of duplicated genes, we now precisely define what is a *matching* between two genomes. Roughly speaking, a matching between two genomes can be seen as a way to describe a putative assignment of orthologous pairs of genes between the two genomes (see for example [11]). A matching $\mathcal{M}$ between genomes G_0 and G_1 is a set of pairwise disjoint pairs $(G_0[i], G_1[j])$, where $G_0[i]$ and $G_1[j]$ belong to the same gene family regardless of the sign, *i.e.*, $|G_0[i]| = |G_1[j]|$. Genes of G_0 and G_1 that belong to a pair of the matching $\mathcal{M}$ are said to be *saturated* by $\mathcal{M}$, or $\mathcal{M}$-saturated for short. A matching $\mathcal{M}$ between G_0 and G_1 is said to be *maximum* if for any gene family, there are no two genes of this family that are unmatched for $\mathcal{M}$ and belong to G_0 and G_1, respectively.

The above definition allows us a large degree of freedom in the choice of the matching between two genomes. Two types of matching are usually considered and specify the underlying model to focus on for computing the desired genomic distance. In the *exemplar* model, the matching $\mathcal{M}$ is required to saturate exactly one gene of each gene family, *i.e.*, the size of the matching is the number of gene families. In the *maximum matching* model, the matching $\mathcal{M}$ is required to saturate as many genes of any gene family as possible, *i.e.*, $\mathcal{M}$ is a matching of maximum cardinality. Let $\mathcal{M}$ be any matching between G_0 and G_1 that fulfills the requirements of a given model (*exemplar* or *maximum matching*). By first deleting non-saturated genes and next renaming genes in G_0 and G_1 according to the matching $\mathcal{M}$, we may now assume that both G_0 and G_1 are duplication-free, *i.e.* G_1 is a signed permutation of G_0. We call the resulting genomes $\mathcal{M}$-*pruned*.

Let G_0 and G_1 be two duplication-free genomes of size n. Without loss of generality, we may assume that G_0 is the identity permutation, *i.e.*, $G_0 = 1\,2\,\ldots\,n$. We say that there is a *breakpoint* after gene $G_0[i]$, $1 \leq i < n$, in G_0 if neither $G_0[i]$ and $G_0[i + 1]$ nor $-G_0[i + 1]$ and $-G_0[i]$ are consecutive genes in G_1, otherwise we say that there is an *adjacency* after gene $G_0[i]$. For example, if $G_0 = 1\,2\,3\,4\,5$ and $G_1 = 1\,-3\,-2\,4\,5$, then we have a breakpoint in G_0 after genes 1 and 3 (and hence we have an adjacency in G_0 after genes 2 and 4).

Let G_0 and G_1 be two genomes and $\mathcal{M}$ be a matching under any model (*exemplar* or *maximum matching*) between G_0 and G_1. We define $A_{\mathcal{M}}(G_0, G_1)$ and $B_{\mathcal{M}}(G_0, G_1)$ to be the number of adjacencies and the number of breakpoints between the two $\mathcal{M}$-pruned genomes.

We are now in position to formally define the optimization problem we are interested in. Given two genomes G_0 and G_1 and a model (*exemplar* or *maximum matching*), find a matching $\mathcal{M}$ between G_0 and G_1 that fulfills the requirements of the model such that the number of breakpoints between the two $\mathcal{M}$-pruned genomes is as small as possible.

3   An Exact Algorithm

3.1   Pseudo-boolean Problem

Minimizing the number of breakpoints between two genomes with duplications is an **NP**–hard problem under the *exemplar* model even when $occ_0(\mathbf{g}) = 1$ for all genes $\mathbf{g}$ in G_0 and $occ_1(\mathbf{g}) \leq 2$ for all genes $\mathbf{g}$ in G_1 [8]. Consequently, the NP-hardness also holds under the *maximum matching* model.

The exact algorithms we define in this section attempt to take advantage of the existing solvers, and more precisely of the linear pseudo-boolean solvers, which are a generalization of the SAT solvers. To this end, we have to express our problem (with its two variants, according to the *exemplar* or *maximum matching* model) as a linear pseudo-boolean problem (or LPB problem), *i.e.* as a linear program [17] whose variables take 0 or 1 values. A number of generalizations of SAT solvers to LPB solvers have been proposed (`Pueblo` [18], `Galena` [9], `OPBDP` [3] and more). We decided to use for our tests the `minisat+` LPB solver [12] because of its good results during PB evaluation 2005 (special track of the SAT COMPETITION 2005).

Instead of directly writing a program that minimizes the number of breakpoints, we chose to write the complementary program which consists in maximizing the number of adjacencies between the two given genomes. There are two reasons for this choice. First, the constraints are simpler and less numerous in this latter case ; moreover, experimental tests moreover showed that the running time of our program is noticeably better by focusing on adjacencies. Second, it is easy to notice that minimizing the number of breakpoints and maximizing the number of adjacencies are equivalent problems under both the *exemplar* and *maximum matching* models. Indeed, according to the above notations, given a matching $\mathcal{M}$ between two genomes G_0 and G_1 we have:

$$B_{\mathcal{M}}(G_0, G_1) + A_{\mathcal{M}}(G_0, G_1) = |\mathcal{M}| - 1. \tag{1}$$

For the *exemplar* and *maximum matching* models, all the matchings $\mathcal{M}$ satisfying the model have the same size, and hence $B_{\mathcal{M}}(G_0, G_1) + A_{\mathcal{M}}(G_0, G_1)$ is a constant. Therefore, maximizing $A_{\mathcal{M}}(G_0, G_1)$ is equivalent to minimizing $B_{\mathcal{M}}(G_0, G_1)$.

3.2   Maximizing the Number of Adjacencies

The `LPB` program we propose considers two genomes with duplications and performs an $\mathcal{M}$-pruning which maximizes the number of adjacencies according to a specified model (*exemplar* or *maximum matching*). As discussed above, the resulting matching also minimizes the number of breakpoints between the two genomes. The `LPB` program, Program `Breakpoint-Maximum-Matching`, for the *maximum matching* model is given in Figure 1. The *exemplar* variant is easily obtained by performing only a few changes that are discussed subsequently.

Program `Breakpoint-Maximum-Matching` considers two genomes G_0 and G_1 of respective lengths n_0 and n_1. The objective function, the variables and the constraints are briefly discussed hereafter.

Program **Breakpoint-Maximum-Matching**

Objective :

Maximize $\displaystyle\sum_{0 \leq i < n_0} \sum_{i < j \leq n_0} \sum_{0 \leq k < n_1} \sum_{k < \ell \leq n_1} d(i, j, k, \ell)$

Constraints :

(C.01) $\forall\, 1 \leq i \leq n_0, \displaystyle\sum_{1 \leq k \leq n_1,\ |G_0[i]|=|G_1[k]|} a(i, k) = b_0(i)$

$\forall\, 1 \leq k \leq n_1, \displaystyle\sum_{1 \leq i \leq n_0,\ |G_0[i]|=|G_1[k]|} a(i, k) = b_1(k)$

(C.02) $\forall\, 0 \leq x \leq 1, \forall\, \mathbf{g} \in \mathcal{G}, \displaystyle\sum_{1 \leq i \leq n_x,\ |G_x[i]|=|g|} b_x(i) = \min(\mathrm{occ}_0(\mathbf{g}), \mathrm{occ}_1(\mathbf{g}))$

(C.03) $\forall\, 0 \leq x \leq 1, \forall\, 1 \leq i \leq j - 1 < n_x, c_x(i, j) + \displaystyle\sum_{i < p < j} b_x(p) \geq 1$

(C.04) $\forall\, 0 \leq x \leq 1, \forall\, 1 \leq i < p < j \leq n_x, c_x(i, j) + b_x(p) \leq 1$

(C.05) $\forall\, 1 \leq i < j \leq n_0, \forall\, 1 \leq k < \ell \leq n_1,$
such that $G_0[i] = G_1[k]$ and $G_0[j] = G_1[\ell]$,
$a(i, k) + a(j, \ell) + c_0(i, j) + c_1(k, \ell) - d(i, j, k, \ell) \leq 3$

(C.06) $\forall\, 1 \leq i < j \leq n_0, \forall\, 1 \leq k < \ell \leq n_1,$
such that $G_0[i] = G_1[k]$ and $G_0[j] = G_1[\ell]$,
$a(i, k) - d(i, j, k, \ell) \geq 0$
$a(j, \ell) - d(i, j, k, \ell) \geq 0$
$c_0(i, j) - d(i, j, k, \ell) \geq 0$
$c_1(k, \ell) - d(i, j, k, \ell) \geq 0$

(C.07) $\forall\, 1 \leq i < j \leq n_0, \forall\, 1 \leq k < \ell \leq n_1,$
such that $G_0[i] = -G_1[\ell]$ and $G_0[j] = -G_1[k]$,
$a(i, \ell) + a(j, k) + c_0(i, j) + c_1(k, \ell) - d(i, j, k, \ell) \leq 3$

(C.08) $\forall\, 1 \leq i < j \leq n_0, \forall\, 1 \leq k < \ell \leq n_1,$
such that $G_0[i] = -G_1[\ell]$ and $G_0[j] = -G_1[k]$,
$a(i, \ell) - d(i, j, k, \ell) \geq 0$
$a(j, k) - d(i, j, k, \ell) \geq 0$
$c_0(i, j) - d(i, j, k, \ell) \geq 0$
$c_1(k, \ell) - d(i, j, k, \ell) \geq 0$

(C.09) $\forall\, 1 \leq i < j \leq n_0, \forall\, 1 \leq k < \ell \leq n_1,$
such that $\{|G_0[i]|, |G_0[j]|\} \neq \{|G_1[k]|, |G_1[\ell]|\}$ or $G_0[i] - G_0[j] \neq G_1[k] - G_1[\ell]$,
$d(i, j, k, \ell) = 0$

(C.10) $\forall\, 1 \leq i < j \leq n_0,$
$\displaystyle\sum_{1 \leq k < n_1} \sum_{k < \ell \leq n_1} d(i, j, k, \ell) \leq 1$

Domains :
$\forall\, x \in \{0, 1\}, \forall\, 1 \leq i < j \leq n_0, \forall\, 1 \leq k < \ell \leq n_1,$
$a(i, k),\ b_x(i),\ c_x(i, k),\ d(i, j, k, \ell) \in \{0, 1\}$

Fig. 1. Program **Breakpoint-Maximum-Matching** for finding the maximum number of adjacencies between two genomes under the *maximum matching* model

Variables

- Variables $a(i,k)$, $1 \leq i \leq n_0$ and $1 \leq k \leq n_1$, define a matching $\mathcal{M}$: $a_{i,k} = 1$ if and only if the gene at position i in G_0 is matched with the gene at position k in G_1 in $\mathcal{M}$.
- Variables $b_x(i)$, $x \in \{0,1\}$ and $1 \leq i \leq n_x$, represent the $\mathcal{M}$-*saturated* genes: $b_x(i) = 1$ if and only if the gene at position i in G_x is saturated by the matching $\mathcal{M}$. Clearly, $\sum_{1 \leq i \leq n_0} b_0(i) = \sum_{1 \leq k \leq n_1} b_1(k)$, and this is precisely the size of the matching $\mathcal{M}$.
- Variables $c_x(i,j)$, $x \in \{0,1\}$ and $1 \leq i < j \leq n_x$, represent *consecutive genes* according to the matching $\mathcal{M}$: $c_x(i,j) = 1$ if and only if the genes at positions i,j in G_x are saturated by $\mathcal{M}$ and no gene at position p, $i < p < j$, is saturated by $\mathcal{M}$.
- Variables $d(i,j,k,\ell)$, $1 \leq i < j \leq n_0$ and $1 \leq k < \ell \leq n_1$, represent *adjacencies* according to the matching $\mathcal{M}$: $d(i,j,k,\ell) = 1$ if and only if (i) either $(G_0[i], G_1[k])$ and $(G_0[j], G_1[\ell])$ are two edges of $\mathcal{M}$, or $(G_0[i], G_1[\ell])$ and $(G_0[j], G_1[k])$ are two edges of $\mathcal{M}$, (ii) $G_0[i]$ and $G_0[j]$ are consecutive in G_0 according to $\mathcal{M}$, (iii) $G_1[k]$ and $G_1[\ell]$ are consecutive in G_1 according to $\mathcal{M}$.

Objective function

The objective of Program `Breakpoint-Maximum-Matching` is to maximize the number of adjacencies between the two considered genomes. This objective reduces in our model to maximizing the sum of all variables $d(i,j,k,\ell)$.

Constraints

Assume $x \in \{0,1\}$, $1 \leq i < j \leq n_0$ and $1 \leq k < \ell \leq n_1$.

- Constraint **(C.01)** ensures that each gene of G_0 and of G_1 is matched at most once, *i.e.*, $b_0(i) = 1$ (resp. $b_1(k) = 1$) if an only if gene i (resp. k) is matched in G_0 (resp. G_1) ; see Figure 2 for an illustration. Moreover, the matching is possible only between genes in the same family. It is worth noticing here that we do not specifically ask that $a(i,k) = 0$ when i and k concern genes belonging to different families. This is simply not necessary.
- Constraint **(C.02)** defines the model (*i.e.* the *maximum matching* model, in this case). For each gene family **g**, one must have a single matched gene for the *exemplar* model and $min(occ_0(\mathbf{g}), occ_1(\mathbf{g}))$ matched genes for the *maximum matching* model (see Figure 2).
- Constraints in **(C.03)** and **(C.04)** express the definition of consecutive genes, thus fixing the values of the variables c_x. The variable $c_x(i,j)$ is equal to 1 if and only if there exists no p such that $i < p < j$ and $b_x(p) = 1$. Again, it is worth noticing that the constraints do not force the variables $c_x(i,j)$ to have exactly the values we intuitively wish according to the abovementioned interpretation. Here, we accept that $c_x(i,j) = 1$ even if the gene at position i or j is *not* matched. However, this will pose no problem in the sequel.
- Constraints in **(C.05)** to **(C.10)** define variables d. In the case where $G_0[i] = G_1[k]$ and $G_0[j] = G_1[\ell]$, constraints **(C.05)** and **(C.06)** ensure that we have

$d(i, j, k, \ell) = 1$ if and only if all variables $a(i, k)$, $a(j, \ell)$, $c_0(i, j)$ and $c_1(k, \ell)$ are equal to 1. In the case where $G_0[i] = -G_1[\ell]$ and $G_0[j] = -G_1[k]$, constraints **(C.07)** and **(C.08)** ensure that we have $d(i, j, k, \ell) = 1$ if and only if all variables $a(i, \ell)$, $a(j, k)$, $c_0(i, j)$ and $c_1(k, \ell)$ are equal to 1. Constraint **(C.09)** fixes the variable $d(i, j, k, \ell)$ to 0 if none of the two cases above holds. Constraint **(C.10)** requires to have at most one adjacency for every pair (i, j). See Figure 3 for a simple illustration.

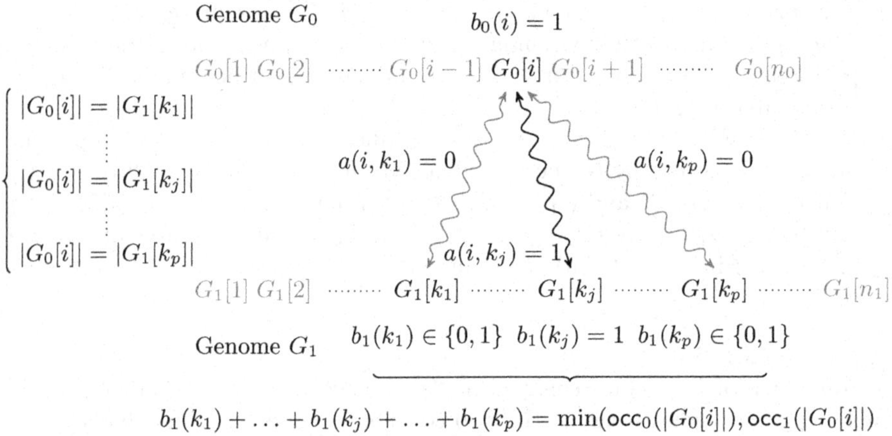

Fig. 2. Illustration of the constraints on variable $b_0(i)$, $1 \le i \le n_0$. If gene $G_0[i]$ appears in positions $k_1 < k_2 < \ldots < k_p$ in G_1 and gene $G_0[i]$ is mapped to gene $G_1[k_j]$ in the solution mapping, then (i) $a(i, k_j) = 1$, *i.e.*, gene $G_0[i]$ is mapped to gene $G_1[k_j]$, (ii) $a(i, k_q) = 0$ for $1 \le q \le p$ and $q \ne j$, *i.e.*, gene $G_0[i]$ is mapped to only one gene in G_1, (iii) $b_0(i) = 1$, *i.e.*, gene $G_0[i]$ is mapped to a gene of G_1 and (iv) $b_1(k_j) = 1$, *i.e.*, gene $G_1[k_j]$ is mapped to a gene of G_0. Observe that one may have in addition $b_1(k_q) = 1$ for some $1 \le q \le p$ and $q \ne j$ if $\min(\mathsf{occ}_0(|G_0[i]|), \mathsf{occ}_1(|G_0[i]|)) \ge 1$ (this observation is however no longer valid for the *exemplar* model).

Program `Breakpoint-Maximum-Matching` has $O((n_0 n_1)^2)$ constraints and $O((n_0 n_1)^2)$ variables, which could result in a time-consuming computation. Several simple rules have been used in order to speed-up the execution, some of which help to reduce the number of variables and constraints. They are discussed in the next subsection.

3.3   Speeding-Up the Program

We briefly describe in this section some rules for speeding-up the pseudo-boolean program.

Pre-processing the genomes. The genomes are pairwise pre-processed to delete all genes that do not appear in both genomes. For the *exemplar* model, consecutive

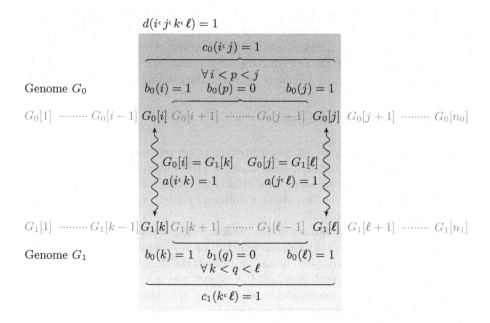

Fig. 3. Illustration of the constraints on variable $d(i, j, k, \ell)$, $1 \leq i < j \leq n_0$ and $1 \leq k < \ell \leq n_1$, for $G_0[i] = G_1[k]$ and $G_0[j] = G_1[\ell]$. The two genes $G_0[i]$ and $G_{[j]}$ are adjacent according to a solution mapping if there exist two genes $G_1[k]$ and $G_1[\ell]$, $G_0[i] = G_1[k]$ and $G_0[j] = G_1[\ell]$, such that (i) $G_0[i]$ is mapped to $G_1[k]$, *i.e.*, $a(i, k) = 1$, (ii) $G_0[j]$ is mapped to $G_1[\ell]$, *i.e.*, $a(j, \ell) = 1$, (iii) no gene between $G_0[i]$ and $G_0[j]$ is mapped to a gene of G_1, *i.e.*, $c_0(i, j) = 1$ and (iv) no gene between $G_1[k]$ and $G_1[\ell]$ is mapped to a gene of G_1, *i.e.*, $c_1(k, \ell) = 1$. The above situation reduces in our modelization to $d(i, j, k, \ell) = 1$.

occurrences of a gene (with the same sign) are reduced to only one occurrence to this gene. For the γ-proteobacteria benchmark set, the average size of a genome reduces from 3000 to 1300.

Reducing the number of variables and constraints. Due to space constraints we only list few easy reduction rules. For non-duplicated genes, *i.e.*, $\mathrm{occ}_0(g) = \mathrm{occ}_1(g) = 1$, the corresponding variable $a_{i,k}$ is set directly to 1, as well as the two variables $b_0(i)$ and $b_1(k)$. Also, if two non-duplicated genes occur consecutively or in reverse order with opposite signs, the corresponding variable $d()$ is set directly to 1 and the related constraints are discarded. For the *exemplar* model, we must have exactly one occurrence of each gene in each genome, and hence if the same gene occurs, say in G_0, at positions i and j, then the corresponding variable $d()$ is set directly to 0 and the related constraints are discarded. If for two genes, say occurring at positions i and j in G_0 and k and ℓ in G_1, at least one gene occurring between position i and j in G_0 or k and ℓ in G_1 must be saturated in any matching $\mathcal{M}$, then the corresponding variable $d(i, j, k, \ell)$ is set directly to 0 and the related constraints are discarded (details omitted).

Adding redundancy. While adding redundancy to a pseudo-boolean program is certainly useless from a correctness point of view, it can however have a major impact on the practical performance of the programs. For example, Program `Breakpoint-Maximum-Matching` contains some redundant constraints (**(C.06)**, **(C.08)** and **(C.10)**) that significantly improved the running time of the program.

4   Experimental Results

Thanks to the `LPB` program discussed previously, as well as formula (1), we are now able to determine the minimum number of breakpoints between pairs of genomes that contain duplicates. This minimum number of breakpoints will be computed according to the two above mentioned models, i.e. the *exemplar* and *maximum matching* models.

To this end, we used a dataset of γ-proteobacteria genomes, originally studied in [13], and exploited several times since then. This dataset is composed of twelve complete linear genomes of γ-Proteobacteria out of the thirteen originally studied in [13]. Indeed, the thirteenth genome (*V.cholerae*) was not considered, since it is composed of two chromosomes, and hence does not fit in the model we considered here for representing genomes. More precisely, the dataset is composed of the genomes of the following species:

- *Buchnera aphidicola APS* (`Baphi`, Genbank accession number NC_002528),
- *Escherichia coli K12* (`Ecoli`, NC_000913),
- *Haemophilus influenzae Rd* (`Haein`, NC_000907),
- *Pseudomonas aeruginosa PA01* (`Paeru`, NC_002516),
- *Pasteurella multocida Pm70* (`Pmult`, NC_002663),
- *Salmonella typhimurium LT2* (`Salty`, NC_003197),
- *Xanthomonas axonopodis pv. citri 306* (`Xaxon`, NC_003919),
- *Xanthomonas campestris* (`Xcamp`, NC_0 03902),
- *Xylella fastidiosa 9a5c* (`Xfast`, NC_002488),
- *Yersinia pestis CO_92* (`Ypest-CO92`, NC_003143),
- *Yersinia pestis KIM5 P12* (`Ypest-KIM`, NC_004088) and
- *Wigglesworthia glossinidia brevipalpis* (`Wglos`, NC_004344).

The computation of a partition of the complete set of genes into gene families, where each family is supposed to represent a group of homologous genes, is taken from [5] (this partition was actually provided to these authors by Lerat [13]). It should be noted that in average, 11% of duplicated genes are present in these genomes.

The `LPB` engine is powered by `minisat+` [12]. Computations were carried out on a Quadri Intel(R) Xeon(TM) CPU 3.00 GHz with 16Gb of memory running under Linux. Under the *maximum matching* model, `minisat+` runs our program `Breakpoint-Maximum-Matching` (implemented using the speeding-up rules described in Section 3.3) in less than 10s for 56 out of the 66 possible pairs of genomes, and in several minutes for the remaining 10 pairs. The results are provided in Table 1.

Table 1. Exact number of breakpoints for the *maximum matching* model

Genomes	Number of Breakpoints (*maximum matching* model)										
Ecoli	156										
Haein	270	665									
Paeru	240	1082	615								
Pmult	259	703	525	681							
Salty	158	277	676	1091	704						
Wglos	170	194	277	260	270	192					
Xaxon	226	842	533	1016	557	854	269				
Xcamp	226	845	530	1012	555	854	268	181			
Xfast	236	564	468	572	481	569	272	400	404		
Ypest-co92	170	596	649	990	671	591	193	760	755	542	
Ypest-kim	176	607	653	1004	676	606	197	760	749	545	59
	Baphi	Ecoli	Haein	Paeru	Pmult	Salty	Wglos	Xaxon	Xcamp	Xfast	Ypest-co92

The first conclusion that can be drawn from these results is the following: the pseudo-boolean approach we have considered here is a good approach for computing the minimum number of breakpoints for the *maximum matching* model, since *all* the results have been obtained within a few minutes. However, as already observed in [1] for maximizing the number of *common intervals* between two genomes, we notice that the *exemplar* model is the main bottleneck of our approach. Indeed, for the *exemplar* model, only 49 out of 66 (that is about 74%) results have been obtained within a few minutes (we stopped the computation of the 17 remaining cases after a few days). We still have no formal explanation for this surprising and counter-intuitive fact. The 49 results we have obtained are given in Table 2.

Besides the fact that computing the minimum number of breakpoints under the *maximum matching* model proves to be feasible under our pseudo-boolean approach, we find interesting to note that we have a sufficient number of results in both the *maximum matching* and the *exemplar* models to test the absolute accuracy of possible heuristics for these two problems. Indeed, if one wishes to obtain fast (though not optimal) results by using a given heuristic, it is relevant to know how tight this heuristic is. We carried out this study, focusing on two heuristics (one for the *maximum matching* model, the other for the *exemplar* model), that are both based on iteratively choosing a Longest Common Substring (LCS).

Maximum Matching Model. In [14], the authors introduced an heuristic that aimed at computing a matching between two genomes. This heuristic is a greedy algorithm based on the notion of *LCS*. Let G_0 and G_1 be two genomes: an *LCS* of (G_0, G_1) is a longest common word S of G_0 and G_1, up to a complete reversal. The idea of the greedy algorithm is to match, at each iteration, all the genes

Table 2. Exact number of breakpoints for the *exemplar* model (49 instances out of 66)

Genomes	Number of Breakpoints (*exemplar* model)										
Ecoli	152										
Haein	265	610									
Paeru	232		550								
Pmult	254	622		592							
Salty	154		612		622						
Wglos	168	183	267	248	262	181					
Xaxon	222	675	473		495	684	261				
Xcamp	222	678	473		495		260				
Xfast	231	491	424	499	436	497	264				
Ypest-co92	166		597		597		182	624	620	473	
Ypest-kim	172		598		601		186	624	618	477	
	Baphi	Ecoli	Haein	Paeru	Pmult	Salty	Wglos	Xaxon	Xcamp	Xfast	Ypest-co92

that are in an *LCS*. If there are several *LCS*, one is chosen arbitrarily. In [1], we improved this heuristic in the following way: at each iteration, not only we match an *LCS*, but we also remove each unmatched gene of a genome, for which there is no unmatched gene of same family in the other genome. These rules imply that the resulting matching is a maximum matching. We call this heuristic IILCS_MM.

Exemplar Model. For the *exemplar* model, we use the same strategy (iteratively match the genes of an *LCS*), except that in this case we must make sure that only *one gene* from each family is matched on each genome. Therefore, at each iteration, and for each gene **g** present in the *LCS* (and thus kept in the matching), we remove all the other occurrences of **g** in both genomes. Let us call this heuristic IILCS_EX.

We have tested both IILCS_MM and IILCS_EX under, respectively, the *maximum matching* and *exemplar* models. Current results are given in Tables 1 to 4 (see http://www.lri.fr/~thevenin/Breakpoint/#Some_results for up-to-date results). The two heuristics are quite fast and one can obtain all results for IILCS_MM and IILC_EX within 20 minutes on a regular desktop computer. For the *maximum matching* model, Heuristic IILCS_MM provides results that are on average 99.11% of the optimal number of breakpoints, ranging from 95.51% to 100%. We actually note that in 14 out of the 66 cases, IILCS_MM returns the optimal value. Concerning IILCS_EX, the average, obtained over the 49 instances for which we know the optimal result, is 96.88%, ranging from 94.38% to 99.10%.

We thus conclude that both heuristics IILCS_MM and IILCS_EX, despite being extremely simple and fast, appear to be very good on the dataset we studied.

Table 3. Number of breakpoints for the *maximum matching* model by IILCS_MM

Genomes	Number of Breakpoints (*maximum matching* model) for Heuristic IILCS_MM										
	Baphi	Ecoli	Haein	Paeru	Pmult	Salty	Wglos	Xaxon	Xcamp	Xfast	Ypest-co92
Ecoli	157										
Haein	270	670									
Paeru	241	1097	619								
Pmult	259	705	529	684							
Salty	158	290	680	1101	708						
Wglos	171	195	277	262	270	193					
Xaxon	226	848	533	1023	560	863	269				
Xcamp	226	851	532	1023	559	860	269	185			
Xfast	236	569	468	575	481	571	272	406	408		
Ypest-co92	173	618	655	1007	678	609	195	767	766	549	
Ypest-kim	178	628	660	1019	684	626	198	766	758	550	59

Table 4. Number of breakpoints for the *exemplar* model by IILCS_EX

Genomes	Number of Breakpoints (*exemplar* model) for Heuristic IILCS_EX										
	Baphi	Ecoli	Haein	Paeru	Pmult	Salty	Wglos	Xaxon	Xcamp	Xfast	Ypest-co92
Ecoli	155										
Haein	268	636									
Paeru	238	888	571								
Pmult	258	657	509	619							
Salty	156	175	641	908	659						
Wglos	170	189	272	254	266	188					
Xaxon	224	712	494	844	516	722	264				
Xcamp	224	716	492	841	516	720	263	126			
Xfast	234	511	443	517	456	514	268	384	383		
Ypest-co92	171	482	619	829	620	491	188	650	648	490	
Ypest-kim	176	485	624	827	623	492	191	649	644	495	34

In particular, for the *exemplar* model, since our pseudo-boolean approach seems to reach its limits for some instances, it could be convenient to compute those remaining instances using Heuristic IILCS_EX.

5 Conclusion

In this paper, we presented a method that helps speeding-up computations of exact results for comparing whole genomes containing duplicates. This method, which makes use of pseudo-boolean programming, has been introduced in [1] for computing the maximum number of common intervals between two genomes, and can be used for several (dis)similarity measures. In this paper, we used this method for computing the minimum number of breakpoints between two genomes, and developed pseudo-boolean programs for both the *maximum matching* and *exemplar* models. Experiments were undertaken on a dataset of γ-Proteobacteria, showing the validity of our approach, since all the results (resp. 49 results out of 66) have been obtained in a limited amount of time in the *maximum matching* model (resp. *exemplar* model). Moreover, these results allow us to state that both the IILCS_MM and the IILCS_EX heuristics provide excellent results on this dataset, hence showing their validity and robustness. On the whole, these preliminary results are very encouraging.

There is still a great amount of work to be done. For instance:

- Implementing and testing the *maximum matching* and the *exemplar* models, for several other (dis)similarity measures,
- For each case, determining strong and relevant rules for speeding-up the process by avoiding the generation of a large number clauses and variables (a pre-processing step that should not be underestimated),
- Obtaining exact results for each of these models and measures, and for different datasets, that could be later used as benchmarks in order to validate (or not) possible heuristics, and
- Implementing and testing an intermediate model between the *maximum matching* and the *exemplar* models, in which one must match *at least* one gene of each family in each genome.

References

1. Angibaud, S., Fertin, G., Rusu, I., Vialette, S.: How pseudo-boolean programming can help genome rearrangement distance computation. In: Bourque, G., El-Mabrouk, N. (eds.) Comparative Genomics. LNCS (LNBI), vol. 4205, pp. 75–86. Springer, Heidelberg (2006)
2. Angibaud, S., Fertin, G., Rusu, I., Vialette, S.: A general framework for computing rearrangement distances between genomes with duplicates. Journal of Computational Biology 14(4), 379–393 (2007)
3. Barth, P.: A Davis-Putnam based enumeration algorithm for linear pseudo-boolean optimization. Technical Report MPI-I-95-2-003, Max Planck Institut Informatik, pages 13 (2005)
4. Blin, G., Chauve, C., Fertin, G.: The breakpoint distance for signed sequences. In: Proc. 1st Algorithms and Computational Methods for Biochemical and Evolutionary Networks (Comp. Bio. Nets.), pp. 3–16. KCL publications (2004)
5. Blin, G., Chauve, C., Fertin, G.: Genes order and phylogenetic reconstruction: Application to γ-proteobacteria. In: McLysaght, A., Huson, D.H. (eds.) RECOMB 2005. LNCS (LNBI), vol. 3678, pp. 11–20. Springer, Heidelberg (2005)

6. Blin, G., Rizzi, R.: Conserved intervals distance computation between non-trivial genomes. In: Wang, L. (ed.) COCOON 2005. LNCS, vol. 3595, pp. 22–31. Springer, Heidelberg (2005)
7. Bourque, G., Yacef, Y., El-Mabrouk, N.: Maximizing synteny blocks to identify ancestral homologs. In: McLysaght, A., Huson, D.H. (eds.) RECOMB 2005. LNCS (LNBI), vol. 3678, pp. 21–35. Springer, Heidelberg (2005)
8. Bryant, D.: The complexity of calculating exemplar distances. In: Comparative Genomics: Empirical and Analytical Approaches to Gene Order Dynamics, Map Alignment, and the Evolution of Gene Families, pp. 207–212. Kluwer Academic Publishers, Dordrecht (2000)
9. Chai, D., Kuehlmann, A.: A fast pseudo-boolean constraint solver. In: Proc. 40th ACM IEEE Conference on Design Automation, pp. 830–835. ACM Press, New York (2003)
10. Chauve, C., Fertin, G., Rizzi, R., Vialette, S.: Genomes containing duplicates are hard to compare. In: Alexandrov, V.N., van Albada, G.D., Sloot, P.M.A., Dongarra, J.J. (eds.) ICCS 2006. LNCS, vol. 3992, pp. 783–790. Springer, Heidelberg (2006)
11. Chen, X., Zheng, J., Fu, Z., Nan, P., Zhong, Y., Lonardi, S., Jiang, T.: Assignment of orthologous genes via genome rearrangement. IEEE/ACM Transactions on Computational Biology and Bioinformatics 2(4), 302–315 (2005)
12. Eén, N., Sörensson, N.: Translating pseudo-boolean constraints into SAT. Journal on Satisfiability, Boolean Modeling and Computation 2, 1–26 (2006)
13. Lerat, E., Daubin, V., Moran, N.A.: From gene tree to organismal phylogeny in prokaryotes: the case of γ-proteobacteria. PLoS Biology 1(1), 101–109 (2003)
14. Marron, M., Swenson, K.M., Moret, B.M.E.: Genomic distances under deletions and insertions. Theoretical Computer Science 325(3), 347–360 (2004)
15. Sankoff, D.: Genome rearrangement with gene families. Bioinformatics 15(11), 909–917 (1999)
16. Sankoff, D., Haque, L.: Power boosts for cluster tests. In: McLysaght, A., Huson, D.H. (eds.) RECOMB 2005. LNCS (LNBI), vol. 3678, pp. 11–20. Springer, Heidelberg (2005)
17. Schrijver, A.: Theory of Linear and Integer Programming. John Wiley and Sons, Chichester (1998)
18. Sheini, H.M., Sakallah, K.A.: Pueblo: A hybrid pseudo-boolean SAT solver. Journal on Satisfiability, Boolean Modeling and Computation 2, 165–189 (2006)

Improving Inversion Median Computation Using Commuting Reversals and Cycle Information

William Arndt and Jijun Tang*

Department of Computer Science and Engineering
University of South Carolina
Columbia, SC 29208, USA
jtang@cse.sc.edu

Abstract. In the past decade, genome rearrangements have attracted increasing attention from both biologists and computer scientists as a new type of data for phylogenetic analysis. Methods for reconstructing phylogeny from genome rearrangements include distance-based methods, MCMC methods and direct optimization methods. The latter, pioneered by Sankoff and extended with the software suite GRAPPA and MGR, is the most accurate approach, but is very limited due to the difficulty of its scoring procedure–it must solve multiple instances of median problem to compute the score of a given tree. The median problem is known to be NP-hard and all existing solvers are extremely slow when the genomes are distant. In this paper, we present a new inversion median heuristic for unichromisomal genomes. The new method works by applying sets of reversals in a batch where all such reversals both commute and do not break the cycle of any other. Our testing using simulated datasets shows that this method is much faster than the leading solver for difficult datasets with only a slight accuracy penalty, yet retains better accuracy than other heuristics with comparable speed. This new method will dramatically increase the speed of current direct optimization methods and enables us to extend the range of their applicability to organellar and small nuclear genomes with more than 50 inversions along each edge. As a further improvement, this new method can very quickly produce reasonable solutions to problems with hundreds of genes.

1 Introduction

Because of the advent of high-throughput sequencing and the consequent reduction in costs, we are seeing an explosion in the amount of genomic data of all types. In particular, the availability of fully sequenced and well annotated genomes allows us to move beyond the mere sequence level in the study of genomic evolution. Once a genome has been annotated to the point where gene homologs can be identified, each gene family can be assigned a unique integer and each chromosome represented by an ordering (a permutation) of signed integers, where the sign indicates the strand. Rearrangement of genes under inversion, transposition, and other operations such as duplications, deletions and insertions, then amount to rearrangements of these orderings. Such rearrangements are known to be an important evolutionary mechanism [11] and their use in reconstructing phylogenies has been studied intensely since the pioneering papers of Sankoff [5,24].

---

* Corresponding author.

G. Tesler and D. Durand (Eds.): RECOMB-CG 2007, LNBI 4751, pp. 30–44, 2007.
© Springer-Verlag Berlin Heidelberg 2007

Biologists have embraced this new source of data in their phylogenetic work [11,19,22] and also in comparative genomics [21], while computer scientists are slowly solving the difficult problems posed by the manipulations of these gene orders [18]. During the past several years, computer scientists have been able to make substantial progress in genome rearrangement research: with the solution for inversion distance [13] and inversion median [10], we were able to estimate phylogenies and ancestral genomes based on inversions (the dominant events in organellar genomes).

There are several widely used methods for genome rearrangement analysis, including neighbor-joining [23], GRAPPA [17], MGR [7] and Badger [15]. Using the later three generally will achieve better accuracy than using distance based methods such as neighbor-joining. The main software packages for reconstructing the inversion (or breakpoint) phylogeny are GRAPPA and MGR. Their basic optimization tool is an algorithm for computing the inversion (or breakpoint) median of three genomes. However, using GRAPPA and MGR to analyze organismal genomes with many events is extremely expensive, because the median computation takes time exponential in both the size of the genomes and the distances among genomes. In this paper, we present a fast yet accurate heuristic using commuting reversals to improve the inversion median computation for both distant and large genomes. We will also provide some discussions regarding inversion medians when the number of events approach saturation.

2 Backgrounds

2.1 Genome Rearrangements

We assume a reference set of n genes $\{g_1, g_2, \cdots, g_n\}$, thus a unichromisomal genome can be represented as a signed ordering of these genes, and each gene is given an orientation that is either positive, written g_i, or negative, written $-g_i$. Genomes can evolve through events including inversions, transpositions and transversions.

Let G be the genome with signed ordering of $g_1, g_2, \cdots, g_n$. An *inversion* between indices i and j $(i \leq j)$, transforms G to a new genome with linear ordering

$$g_1, g_2, \cdots, g_{i-1}, -g_j, -g_{j-1}, \cdots, -g_i, g_{j+1}, \cdots, g_n$$

A *transposition* on genome G acts on three indices i, j, k, with $i \leq j$ and $k \notin [i, j]$, picking up the interval $g_i, g_{i+1}, \cdots, g_j$ and inserting it immediately after g_k. Thus genome G is replaced by (assume $k > j$):

$$g_1, \cdots, g_{i-1}, g_{j+1}, \cdots, g_k, g_i, g_{i+1}, \cdots, g_j, g_{k+1}, \cdots, g_n$$

An *transversion* is a transposition followed by an inversion of the transposed subsequence; it is also called an *inverted transposition*.

2.2 Distance Computation

Given two genomes G_1 and G_2, we define the *edit distance* $d(G_1, G_2)$ as the minimum number of events required to transform one genome into the other. The *breakpoint distance* [24] is not a direct evolutionary distance measurement. A breakpoint in G_1 is defined as an ordered pair of genes (g_i, g_j) such that g_i and g_j is adjacent in G_1 but not in G_2. The breakpoint distance is simply the number of breakpoints in G_1 relative to G_2. When only inversions are allowed, the edit distance is the *inversion distance*. Hannenhalli and Pevzner [13] developed a mathematical and computational framework for signed gene-orders and provided a polynomial-time algorithm to compute the edit distance between two signed gene-orders under inversions; Bader et al. [1] later showed that this edit distance can be computed in linear time. However, computing the inversion distance is NP-hard in the unsigned case [9].

The HP algorithm is based on the breakpoint graph (Fig 1). We assume without loss of generality that one permutation is the identity. We represent gene i by two vertices, $-i$ and $+i$, connected by an edge. The edge is oriented from $+i$ to $-i$ when gene i is positive, but oriented in the reverse direction when it is negative. Two additional vertices 0 and $n+1$ are also added. These vertices can be connected with two sets of edges, one for each genome. One set of edges, called desire edges, represent the identity genome and is shown with dashed arcs in Fig 1. The other edges represent the current state of the genome and are shown with solid lines; these are called reality edges. In both set of edges, 0 always connects to -1 and $n+1$ always connects to $+n$. The crucial concept is that of alternating cycles in this graph, i.e., cycles of even length in which every odd edge is a desire edge and every even one is a reality edge. Overlapping cycles in certain configurations create structures known as hurdles, and a very unlikely configuration of such hurdles can form a single fortress (please refer [13] for details). Hannenhalli and Pevzner [13] proved that the inversion distance between two signed permutations of n genes is given by

$$n - \# \text{ cycles} + \# \text{ hurdles} + (1 \text{ if fortress present, 0 otherwise})$$

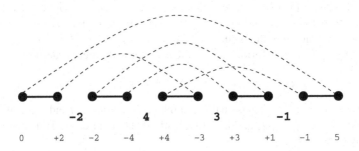

Fig. 1. Breakpoint graph between genome (-2 4 3 -1) and the identity genome (1 2 3 4)

2.3   Sorting and Commuting Reversals

Almost all the distance computation methods return *one* (and only one) minimum sorting sequence. Siepel [26] extended the HP theorem to find *all sorting reversals*, i.e., all possible inversions that appear as the first step in the sorting. Fig. 2 gives one example of sorting reversals: there are eight possible inversions that bring G_2 one step closer to G_1 (the identity genome). This algorithm can be easily extended to enumerate all

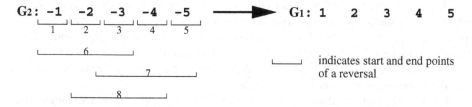

Fig. 2. Eight sorting reversals that bring (-1 -2 -3 -4 -5) one step closer to (1 2 3 4 5)

minimum sorting sequences by identifying every sorting reversals at each step of the sorting.

The concept of commuting reversals was introduced in [6] where the context was sorting between two permutations using inversions. Two different sorting inversions acting on the same permutation are defined to commute as follows: separate the permutation integers into three sets: those integers which are only members of the first inversion, those which are only members of the second, and those which are members of both. The two sorting inversions commute if and only if one of these three sets is empty. Commuting inversions have a desirable property that applying them to a permutation will always obtain the same result no matter the order they are applied.

1 ⌐2 3⌐ 4 5 6	1 ⌐2 3 4 5⌐ 6	1 ⌐2 3 4⌐ 5 6
A	A	A
B	B	B
(a)	(b)	(c)
set only affected by A: [2, 3]	set only affected by A: [5]	set only affected by A: [2]
set only affected by B: [4, 5]	set only affected by B: ϕ	set only affected by B: [5]
set affected by A ∩ B: ϕ	set affected by A ∩ B: [2, 3, 4]	set affected by A ∩ B: [3, 4]

Fig. 3. Examples of commuting inversions (a and b) and non-commuting inversion (c)

2.4 Inversion Median Problem

Given three genomes (permutations) π_1, π_2, π_3 and another genome π_0, we define the *median score* from π_0 as $d(\pi_0,\pi_1) + d(\pi_0,\pi_2) + d(\pi_0,\pi_3)$. The median problem on three genomes is to find genome π_0 that minimizes the median score between itself and each of the three given genomes π_1, π_2, π_3. We also define the *perfect median score* as $\left\lceil \frac{d(\pi_1,\pi_2)+d(\pi_1,\pi_3)+d(\pi_2,\pi_3)}{2} \right\rceil$, which is the lower bound of the score for a median problem.

The median problem is NP-hard [8,20] even for simple distance definition such as breakpoint distance. Seeking a median that minimizes the breakpoint distance can be transformed into a special instance of the well-studied Traveling Salesperson Problem [4], hence can be solved relatively efficiently. But in practice, the breakpoint median is not effective—it is easy to obtain trivial solutions (where the median gene-order coincides with one of the leaves), hence is not as accurate as using inversion median for genome rearrangement analysis [16].

The *inversion median* problem is to find a median genome that minimizes the sum of inversion distances on the three edges. Four inversion median solvers have been proposed. Caprara's solver [10] is based on an extension of the breakpoint graph, while that developed by Siepel and Moret [25] runs a direct search. Both Caprara's and Siepel's median solvers are exact and are included in GRAPPA. In practice, Caprara's median solver is faster when the genomes are not close.

Both MGR [7] and rEvoluzer [2] are heuristics, using similar approach: they both seek good reversals that bring a genome closer to the ancestral genome. For three genomes, the MGR algorithm evaluates all possible inversions for each of the three genomes, identifying good reversals that bring a genome closer to the ancestral genome. Since the ancestral genome is unknown, the algorithm chooses inversions which make G_1 closer to both G_2 and G_3 as *good reversals*. Thus the algorithm will iteratively carry on good reversals in the three genomes until all three are transformed into an identical genome, which is viewed as the most likely ancestral median. rEvoluzer improves the MGR procedure by selecting inversions that cannot destroy any conserved intervals. Although rEvoluzer achieves some speedup over GRAPPA, like MGR, the median gene orders it obtained is not as good as those returned by GRAPPA [2].

All these median solvers become extremely slow for large and distant genomes. A common speedup process used by all methods makes use of the concept of conserved adjacency. A gene pair (x, y) is conserved adjacent if (x, y) or its inverse $(-y, -x)$ is present in all genomes as consecutive elements [14]. A block of k adjacent genes can be replaced by a new gene and the total number of genes reduces by $k - 1$ [7]. This condensation procedure is very effective when the genomes are close: a median of genomes with $1,000$ genes and 50 inversions per edge can be condensed to ~ 200 genes only. In practice, given the smallest edge length e and number of genes n, we found the ratio $\frac{e}{n}$ is a good indicator about the difficulty of inversion median problem. Siepel's median solver cannot handle datasets with $\frac{e}{n} > 15\%$, and its search approach limits it to small genomes (< 100 genes) as well. On the other hand, Caprara's median solver will be able to handle datasets with $1,000$ genes for $\frac{e}{n} \leq 20\%$.

3   Inversion Median Computation Using Commuting Reversals

We set out to improve the speed of inversion median computation with the goal that the new median solver should have accuracy that is comparable to Caprara's median solver. The new algorithm is different from MGR and rEvoluzer in that it will conduct a direct search from one of the known genomes, using sorting reversals to limit the search space. Our algorithm also improves over Siepel's method by using commuting reversals in the set of sorting reversals from the start genome to both of the other two genomes. Our new median solver will also report multiple solutions, a property lacking in almost all existing methods.

3.1   A Naive Approach

Let us first present a naive approach. Suppose the three input permutations are π_1, π_2, and π_3, and assume all median scores are with respect to π_1, π_2, and π_3. Define a recursive function which has input π_1, π_2, π_3, and π_4, where π_4 is set as π_1 when the function is first called. In this function first obtain two sets of sorting reversals, set α which contains sorting reversals from π_4 to π_2, and set β which contains sorting reversals from π_4 to π_3. Let set γ be the intersection of α and β. Repeat the following process until γ is empty: remove one inversion from γ and apply to π_4 to obtain π_4', determine the median score of π_4' and compare to the lowest median score seen so far in the search. If the median score of π_4' is less than or equal to the best-so-far, report the score and π_4'. Call the recursive function with arguments π_1, π_2, π_3, and π_4'.

Several concerns make this method undesirable. Primarily, the amount of computation required increases exponentially with both the number of inversions separating the three permutations and the number of genes in each genome. Second, it can be shown by exhaustively searching permutations against a small inversion median problem, that a median permutation does not necessarily lie on a sorting path between two of the three initial permutations; thus the presented naive approach cannot guarantee an optimal solution because some and possibly all paths to medians would require that one or more reversals which are not members of γ be chosen. We will not attempt to improve this aspect of the naive method as doing so would require a large number of additional inversions be considered in set γ with very little return on the massive amount of new computation being performed.

The biggest problem of the naive approach is that it performs a large amount of redundant computation by visiting the same permutation multiple times. This can be reduced by using information about commuting reversals. Imagine a set of sorting reversals which sort π_1 towards both π_2 and π_3. Select any pair of these reversals A and B which occur along the path to a median. if reversals A and B do not commute, then changing the order that A and B are applied affects the resulting permutation (Fig 3c); if A and B commute (Fig 3a and Fig 3b) then the naive method will search the same permutation at least twice, since both choices of ordering the application of A and B result in the same permutation.

3.2   An Improved Algorithm

The above analysis leads to a method to speed up the search by removing a large portion of this redundancy. Obtain from the set γ reversals with the additional property that all pairs of inversions commute. This allows the order of applying these reversals to be ignored; every permutation which can be reached by applying any number of these commuting reversals can be enumerated and scored one time instead of enumerating permutations by the paths which lead to them. If n is the number of commuting reversals, then 2^n permutations can be reached, but the total number of paths to these permutations is $O(n^n)$.

Which set of the 2^n permutations should be chosen? We have experimented with several methods:

- Brute force method which scores every 2^n permutation and chooses the best median score π_4', with good results but an obvious time complexity drawback.
- A method which draws samples from the 2^n permutations and chooses the best median score among them. This approach reduces both the time required and the

accuracy; in general the quality of the results are proportional to the fraction of the space being searched.

- The simplest method of all, and surprisingly effective, is to apply all reversals in the set, i.e., obtaining π_4' from by applying all non-interfering reversals to π_4. The quality of this method depends on the ratio of the size of the permutation to the size of the reversal set. This approach works well until the $\frac{e}{n}$ ratio is $30-40\%$. Beyond that point each search step normally increases the median score and tends to converge with worse results than a trivial solution.

The previous example, where applying all commuting reversals results in a worse median score, demonstrates there is a more complex interaction between the application of a single sorting reversal to a permutation and its influence on other sorting reversals. This interference between sorting reversals comes from the breaking of cycles in the breakpoint graphs of the problem instance. Imagine a breakpoint graph with one cycle containing two sorting reversals that commute. Applying either of those sorting reversals will alter the breakpoint graph to create two cycles. Afterwards, two possibilities exist: either both of the reality edges of the second reversal will remain in the same cycle, in which case this reversal will be a sorting reversal, or the reality edges of the second reversal will be separated into different cycles, in which case it will no longer be a sorting reversal. This line of thought leads to a concept similar to commuting reversals, only transferred to breakpoint graph cycle interactions.

3.3 Parallel and Perpendicular Sorting Reversals

We call a pair of sorting reversals *parallel* on a single breakpoint graph if they commute, break reality edges in the same cycle of the breakpoint graph, and applying both inversions to the permutation creates two additional cycles. On the other hand, a pair of sorting reversals are *perpendicular* on a single breakpoint graph if they commute, break reality edges in the same cycle of the breakpoint graph, and applying both inversions to the permutation creates one additional cycle.

When multiple breakpoint graphs present, we also call a pair of reversals parallel if in all such graphs the reality permutation is the same (the generalization that the desire permutation is the identity is relaxed), both inversions sort each graph, and the inversions are not perpendicular on any of the considered graphs. A pair of reversals are perpendicular over multiple breakpoint graphs if in all such graphs the reality permutation is the same, the inversions sort each graph, and the inversions are perpendicular on any of the graphs.

Theorem 1. *Consider the breakpoint graph from π_4 to π_2, the breakpoint graph from π_4 to π_3 and the set of commuting sorting inversions γ. Inversions can be removed from the set until no two inversions are perpendicular over both graphs, and the result is a set of sorting inversions which when applied to π_4 in any order improve its median score by an amount equal to the number of inversions which remain in γ.*

Proof. Given set γ contains inversions which sort π_4 towards both π_2 and π_3, that no pair of these inversions are perpendicular, and that all pairs of inversions commute. We prove the theorem by induction. As the inductive step, repeatedly remove any inversion from

γ and apply it to π_4. No pair of inversions in γ are perpendicular, so no previous choice was possible which will affect the sorting property of the current inversion. The chosen inversion is a sorting inversion to both targets, so $d(\pi_4,\pi_1)$ will increase by 1, $d(\pi_4,\pi_2)$ and $d(\pi_4,\pi_3)$ will both decrease by 1. During one inductive step the median score will improve by 1 and the cardinality of γ will lower by one. In the base case, set γ is empty and there are no additional inversions which will sort π_4 towards both π_2 and π_3.

Fig. 4 describes a simple graph method to visualize the parallel and perpendicular inversion properties. We first obtain the set of sorting inversions between two permutations, but additionally save the cycle membership and order in which each reality edge appears when transversing a cycle. For each cycle, in a ring, draw a break location node for each reality edge in the breakpoint graph and label the node with the genes that appear on each side of the edge, and draw an edge representing the desire edge to both of its neighbors. For each sorting reversal which acts on two reality edges in the same cycle (not inversions which merge hurdles or cut a hurdle or fortress), draw a cut chord connecting both of the corresponding break location nodes in the ring. This chord corresponds to the cut that divides the cycle into two smaller cycles when the inversion is applied, and shows which break location nodes will remain in the same cycle and which will be separated. For every pair of inversions in the same cycle, if the cut chord for each intersect, then this pair of inversions is perpendicular, otherwise, the inversions are parallel.

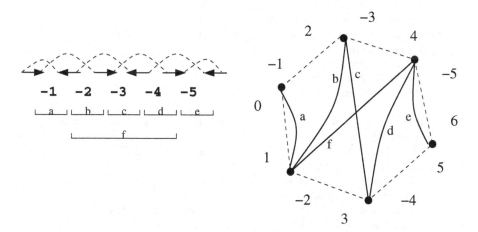

Fig. 4. Graph representation of cycle interferences on a set of commuting reversals, genes 0 and 6 represent linear chromosome endpoints

A special case exists where two inversions share a break location node, as sometimes such inversions will be parallel and sometimes perpendicular. The problem is that when an inversion is applied it separates the two genes labeling a break location node and puts one in each of the newly formed cycles, but different inversions make this decision in different ways, depending on the layout of the permutation. We address this issue in the implementation by treating any inversions which share a break location as perpendicular, with the drawback of sometimes removing inversions from γ which do not actually need to be removed.

3.4   The Final Algorithm

The overall heuristic of our new unichromosomal median solver is presented in Fig 5. Several details worth mentioning. First, the choice of the start permutation has some impact and our experiments show that using the permutation nearest to the center produces the best median scores. Second, despite our efforts to prevent redundant computation, a very large amount still occurs and we used a permutation hash table to check for redundant search paths. This is not a critical aspect and can be removed with little impact–in fact, due to memory constraints it must be removed for genomes larger than approximately 400 genes. The last to mention is the use of set Δ. This is inherited from the naive method as a way to allow every sorting reversal at least one chance to be searched. We will continue to investigate better ways of directing the search past the initial steps.

InversionMedianSolver: input permutations π_1, π_2, π_3
    Compute the pairwise inversion distances between π_1, π_2, π_3
    Choose the one with the smallest sum of its two distances as π_4
    If π_1 was not assigned to π_4, swap it with the permutation which was
    Initilize a global variable *BestSoFar* to an arbitrarily large value
    Call *RecursiveSearch*: π_1, π_2, π_3, π_4

RecursiveSearch: input permutations π_1, π_2, π_3, π_4
    Obtain set α of sorting inversions from π_4, π_2
    Obtain set β of sorting inversions from π_4, π_3
    Obtain set γ, the intersection of α and β
    Call *UseGamma*: γ, π_1, π_2, π_3, π_4
    While set Δ contains elements:
        Set γ to Δ and clear Δ
        Call *UseGamma*: γ, π_1, π_2, π_3, π_4

UseGamma: input set γ and permutations π_1, π_2, π_3, π_4, output set Δ
    For each pair of inversions in γ:
        If inversions do not commute add 1 weight to each;
    Repeat until the weight of all inversions in γ is 0:
        Find the inversion A with largest weight
        Remove A from γ and place in set Δ
        Reduce weight of each inversion not commute with A by 1
    For each pair of inversions in γ:
        If pair is perpendicular, add 1 weight to each
    Repeat until the weight of all inversions in γ is 0:
        Find the inversion A with largest weight and remove A from γ
        Reduce weight of each inversion perpendicular to A by 1
    Apply the reversals in set γ to π_4 to create π_4'
    Calculate the median score of π_4' with respect to π_1, π_2, π_3
    If the score is less than or equal to *BestSoFar*
        Assign the score to *BestSoFar* and output π_4'
    If the median score of π_4' is less than the score of π_4
        Call *RecursiveSearch*: π_1, π_2, π_3, π_4

Fig. 5. Algorithm overview for the new inversion median solver

4   Experimental Results

4.1   Setup of Simulations

We set out to examine the performance (in terms of speed and accuracy) of the new method, using simulated datasets. Because all existing median solvers have very good performance when genomes are close, we only test distant genomes and compare our method against Caprara's solver (slower but exact), and MGR (faster but less accurate).

We focused our experiments on organelle genomes and generated datasets of three genomes with 100 genes for each genome (larger genome sizes were also tested). We first generated trees with three leaves and one internal node, assigned the identity permutation on the internal node and generate the three leaves by applying rearrangement events along each edge respectively. The number of events on each edge is governed by two parameters: the number of overall evolutionary events and the tree shape. We used various number of evolutionary rates: letting r denote the total number of events along all three edges, we used values of r in the range of 80 to 140. We found from our experience that the tree shape plays an important role in median computation, thus we used three tree shapes for each r: a tree with almost equal length edges, i.e., the ratio of three edges are $(1 : 1 : 1)$; a tree with one edge a bit longer than the other two, i.e., of ratio $(2 : 1 : 1)$; a tree with on edge much longer than the other two, i.e., of ratio $(3 : 1 : 1)$. While all computations were based on inversion distances and inversion medians, we generated the data with a deliberate model mismatch to test the robustness of the methods, using a mix of 80% inversions and 20% transpositions. For each combination of parameter settings, we ran 10 datasets and averaged the results. Experiments were conducted on a Linux cluster with 152 Intel Xeon CPUs, but each CPU works independently on a test task. MGR command line options -c -H1 were used.

4.2   Accuracy

Caprara's median solver had no problem to finish all datasets with evolutionary rate $r = 80$ and $r = 100$; however, it could finish a very small number of datasets for $r = 120$ and 140: only four out of 60 datasets finished within 48 hours of computation. Here we report the result separately using slightly different criteria for $r \leq 100$ and $r \geq 120$.

For $r \leq 100$, we report the average median score from our method, Caprara's solver, and MGR. We also report the average perfect (lower bound) median score, which is the best possible score for any median solver. Table 1 shows the result, which indicates that our method is very accurate, with $< 1.5\%$ errors. Our method is most accurate when all three edges are almost equal length, with 70% datasets report median score to those found by Caprara's, while the other 30% are only one score away.

For $r \geq 120$, since Caprara's solver cannot finish all datasets, we only report the average median score from our method and compare it to MGR and the average perfect median score. Table 2 shows the result, which indicates that our method can find better medians than MGR does.

An additional measure of the quality of results is the distance to the simulated ancestor genome. Here we report the average distance to the ancestor for the three methods for those Caprara's method could complete in Table 3, and in Table 4 those sets which Caprara's method could not solve.

Table 1. Comparison of median scores for $r \leq 100$

	(1:1:1)		(2:1:1)		(3:1:1)	
	r=80	r=100	r=80	r=100	r=80	r=100
Prefect median score	86.2	104.2	89.4	105.8	85.7	101.3
Caprara's median score	87.9	107.6	91.4	109.8	88.0	105.2
New method's median score	88.2	109.5	91.8	111.4	89.1	106.7
MGR median score	90.3	113.7	94.3	116.8	89.8	110

Table 2. Comparison of median scores for $r \geq 120$

	(1:1:1)		(2:1:1)		(3:1:1)	
	r=120	r=140	r=120	r=140	r=120	r=140
Prefect median score	116.1	123.5	116.1	122.7	110.3	117.6
New method's median score	125.8	135.3	124.5	134.7	117.9	127.0
MGR median score	132.9	143.6	131.4	142.8	123.6	135.1

Table 3. Comparison of distance to simulated ancestor for $r \leq 100$.

	(1:1:1)		(2:1:1)		(3:1:1)	
	r=80	r=100	r=80	r=100	r=80	r=100
Caprara's method	4.5	18.2	7.2	16.8	5.2	15.7
New Method	9.3	21.7	9.6	20.4	7.2	18.2
MGR	9.3	23.5	11	25.4	9.1	18.6

Table 4. Comparison of distance to simulated ancestor for $r \geq 120$

	(1:1:1)		(2:1:1)		(3:1:1)	
	r=120	r=140	r=120	r=140	r=120	r=140
New method	39.8	49.3	35.2	45.4	23.1	32.1
MGR	40.7	51.6	37.5	49.5	29.7	37.7

4.3   Speed

We recorded the running time for each run as well. Since our method will report all results it can find, there are two measures: 1) the time it finds the first result, and 2) the average number of results it finds within the limit of one hour.

Table 5 and Table 6 shows the first time comparison. When the datasets are relatively easy ($r = 80$), Caprara's solver is much faster than our method. However, it slows down very quickly when the difficulty increases, and almost no dataset can be finished for $r \geq 120$. Meanwhile, the running time of our method is quite consistent with fewer than 30 minutes were used even for the most difficult datasets, which is comparable to the speed of MGR.

In general, our method found 12 medians with the same score within one hour. However, the number is not consistent: some datasets have only one result, while others

Table 5. Comparison of running time for $r \leq 100$ (in seconds)

	(1:1:1)		(2:1:1)		(3:1:1)	
	r=80	r=100	r=80	r=100	r=80	r=100
New method's time	324	551	123	409	1.6	9.3
MGR time	11.2	51.9	11.6	78.2	10.3	35
Caprara's time	3.6	12876	57.2	31387	4.3	6908

Table 6. Comparison of running time for $r \geq 120$ (in seconds)

	(1:1:1)		(2:1:1)		(3:1:1)	
	r=120	r=140	r=120	r=140	r=120	r=140
New method's time	1485	1187	673	453	30	226
MGR time	271.6	560.1	237.8	626.9	135.3	385.4
Caprara's time	> 172880	> 172880	> 172880	> 172880	> 172880	> 172880

Table 7. Average number of medians found for each test case

	r=80	r=100	r=120	r=140
(1:1:1)	27.4	4.0	29.9	7.1
(2:1:1)	11.6	22.9	13.8	8.9
(3:1:1)	6.0	9.1	5.6	11.4

have as many as 120 results. Additionally, by checking inversions on the found medians which do not change the median score, on average for each found median two more
can be quickly located, though they are not significantly different from those already
found. Table 7 shows the average number of medians found by our method.

4.4 Medians of Larger Genomes

We tested the performance of our new method on some simulated datasets of larger
genomes. The simulations were created with the same parameters, except the number
of genes was increased to 500. The tested trees all have edge lengths of 100, 100, and
200, producing a tree with $\frac{e}{n} = 20\%$. Neither Caprara's solver nor MGR could produce
results for any of these trees. Our method, however, can consistently return medians
which are within 7% of the lower bound in less than 30 seconds.

5 Discussion

The method presented here offers acceptable solutions for the median problem which
previous methods are either completely unable to solve in a timely manner, or less accurate. The source of speedup is that our search complexity increases not directly with the
number of genes and events, but with the size of the set of sorting reversals which sort
one input permutation towards both of the others. The two types of problems which our
method performs relatively poorly are those with a small number of events, which limit

the set of shared reversals causing the search to be too shallow, and problems that have all three edges of approximately equal length causing the shared sorting reversal set to be too large resulting in an overly broad search. Real world problems, however, can either be limited to small data sets, in which case existing tools are well sufficient, or may require consideration of large genomes of which such genomes are rarely symmetrical, making our method appropriate.

We believe there is a big problem in the general approach of using the inversion median problem to solve phylogenetic trees composed of distant genomes, a topic discussed in detail in [12]. The direct optimization methods (GRAPPA and MGR) are based upon minimizing the number of inversion events, which requires either the false assumption that there is only one optimal median solution for a given problem instance, or the slightly weaker assumption that although multiple optimal solutions exist, they are all equally valuable for construction of trees.

The issue is most evident when viewing the distances between found medians and the true ancestor: although it is still proportional to median score, increasing the number of events causes the optimal median and ancestor genome to diverge. Several of our test simulations demonstrate the existence of multiple medians that form trees with edges differing by 30% or more, although the tree scores are equal. This shows instances of a median problem do not contain the amount of exact information which current tree methods presume they do. We do not believe that this cause is hopeless however, instead, the notion that any median with optimal score is an equal representative of an internal node should be replaced; new methods or tree building algorithms should be devised which use multiple medians as an intermediate step moving closer to the true ancestor. To obtain more accurate results for large genomes, we may need to find as many medians as possible and choose the one with the minimal total distance to all the others as the representative, or we may need to consider permutations with slightly less than optimal score if they appear in sufficiently large clusters.

The structure of median problems has further vulnerabilities. We ran a small experiment where a random median problem of 10 genes was created, and the median score of every permutation ($2^{10}10!$) in an exhaustive search was found. There are several findings from this experiment: 1) as confirmed by other researchers [3], there exist multiple medians—we found 81 medians for this experiment, with median score of 15; 2) Some of the medians were as far as 9 inversions from one another, but all these 81 medians gathered together in a cloud at the center of the problem space. In order to transverse from one of the initial permutations to a median, the choices of inversions remain very limited, and the possible paths remain very close, until the search nears where medians are located. If this general structure is similar to that of a large median problem instance, then this could be how using commuting reversals achieves its speedup, and additional methods to exploit this structure would likely exist.

6   Conclusions and Future Work

In this paper we present a new inversion median solver using commuting reversals, and introduced the concept of parallel and perpendicular sorting reversals. We extensively tested the method and compare its performance with the leading median solver, using simulated datasets. The experimental results showed that our method is very accurate, and is much faster than the leading solver when the datasets are difficult. We will test the effectiveness of this median solver for phylogenetic reconstruction, and develop methods that can effectively use the multiple median solutions returned by our method to get closer to the true ancestor. We will also extend the concept of solving medians based on breakpoint cycle interactions. If care is taken when hurdles or a fortress are present, the visualization of breakpoint graph cycle interactions could be used to find an upper bound on sorting inversions by predicting and tracking the behavior as the number of cycles increases. This would proceed by examining which and how many potential reversals cut the cycles of other sorting inversions, and thus choose a inversion to apply which separates the break locations of least number of other possible sorting inversions.

Acknowledgments

The authors were supported by US National Institutes of Health (NIH grant number R01 GM078991-01) and by the University of South Carolina. WA is also supported by a fellowship provided by Rothberg foundation. We also thank the anonymous reviewers for their comments and suggestions of improving the original draft.

References

1. Bader, D.A., Moret, B.M.E., Yan, M.: A fast linear-time algorithm for inversion distance with an experimental comparison. J. Comput. Biol. 8(5), 483–491 (2001)
2. Bernt, M., Merkle, D., Middendorf, M.: Genome rearrangement based on reversals that preserve conserved intervals. IEEE/ACM Trans. on Comput. Biol. and Bioinfo. 3(3), 275–288 (2006)
3. Bernt, M., Merkle, D., Middendorf, M.: Using median sets for inferring phylogenetic trees. Bioinformatics 23(2), 129–135 (2007)
4. Blanchette, M., Sankoff, D.: The median problem for breakpoints in comparative genomics. In: Jiang, T., Lee, D.T. (eds.) COCOON 1997. LNCS, vol. 1276, pp. 251–263. Springer, Heidelberg (1997)
5. Blanchette, M., Bourque, G., Sankoff, D.: Breakpoint phylogenies. In: Miyano, S., Takagi, T. (eds.) Genome Informatics, pp. 25–34. Univ. Academy Press (1997)
6. Bergeron, A., Chauve, C., Hartman, T., St-Onge, K.: On the Properties of Sequences of Reversals that Sort a Signed Permutation. Journal Ouvertes Biologie, Informatique, Mathematiques (JOBIM 2002), 99–108 (2002)
7. Bourque, G., Pevzner, P.: Genome-scale evolution: Reconstructing gene orders in the ancestral species. Genome Research 12, 26–36 (2002)
8. Caprara, A.: Formulations and hardness of multiple sorting by reversals. In: Proc. 3rd Int'l Conf. on Comput. Mol. Biol. RECOMB99, pp. 84–93. ACM Press, New York (1999)
9. Caprara, A.: Sorting permutations by reversals and Eulerian cycle decompositions. SIAM J. Discrete Math. 12(1), 91–110 (1999)
10. Caprara, A.: On the practical solution of the reversal median problem. In: Gascuel, O., Moret, B.M.E. (eds.) WABI 2001. LNCS, vol. 2149, pp. 238–251. Springer, Heidelberg (2001)
11. Downie, S., Palmer, J.: Use of chloroplast DNA rearrangements in reconstructing plant phylogeny. In: Soltis, P., et al. (eds.) Plant Molecular Systematics, pp. 14–35 (1992)

12. Eriksen, N.: Reversal and Transposition Medians. Theoretical Computer Sicence 374, 111–126 (2007)
13. Hannenhalli, S., Pevzner, P.A.: Transforming cabbage into turnip (polynomial algorithm for sorting signed permutations by reversals). In: Proc. 27th Ann. Symp. Theory of Computing STOC'95, pp. 178–189. ACM Press, New York (1995)
14. Hannenhalli, S., Pevzner, P.A.: To cut... or not to cut (applications of comparative physical maps in molecular evolution). In: Proc. 7th ACM-SIAM Symp. on Discrete Algorithms (SODA'96), pp. 304–313. SIAM Press (1996)
15. Larget, B., Simon, D.L., Kadane, J.B., Sweet, D.: A Bayesian analysis of metazoan mitochondrial genome arrangements. Mol. Biol. and Evol. 22(3), 486–495 (2005)
16. Moret, B.M.E., Siepel, A., Tang, J., Liu, T.: Inversion medians outperform breakpoint medians in phylogeny reconstruction from gene-order data. In: Guigó, R., Gusfield, D. (eds.) WABI 2002. LNCS, vol. 2452, pp. 521–536. Springer, Heidelberg (2002)
17. Moret, B.M.E., Wyman, S., Bader, D.A., Warnow, T., Yan, M.: A new implementation and detailed study of breakpoint analysis. In: Proc. 6th Pacific Symp. on Biocomputing (PSB 01), pp. 583–594. World Scientific Pub, Singapore (2001)
18. Moret, B.M.E., Tang, J., Warnow, T.: Reconstructing phylogenies from gene-content and gene-order data. In: Gascuel, O. (ed.) Mathematics of Evolution and Phylogeny, pp. 321–352. Oxford Univ. Press, Oxford, UK (2005)
19. Palmer, J.: Chloroplast and mitochondria genome evolution in land plants. In: Herrmann, R. (ed.) Cell Organelles, pp. 99–133 (1992)
20. Pe'er, I., Shamir, R.: The median problems for breakpoints are NP-complete. Elec. Colloq. on Comput. Complexity 71 (1998)
21. Pevzner, P., Tesler, G.: Human and mouse genomic sequences reveal extensive breakpoint reuse in mammalian evolution. Proc. of Natl. Acad. of Sci. USA 100, 7672–7677 (2003)
22. Raubeson, L., Jansen, R.: Chloroplast DNA evidence on the ancient evolutionary split in vascular land plants. Science 255, 1697–1699 (1992)
23. Saitou, N., Nei, M.: The neighbor-joining method: A new method for reconstructing phylogenetic trees. Mol. Biol. Evol. 4, 406–425 (1987)
24. Sankoff, D., Blanchette, M.: Multiple genome rearrangement and breakpoint phylogeny. J. Comput. Biol. 5, 555–570 (1998)
25. Siepel, A., Moret, B.M.E.: Finding an optimal inversion median: experimental results. In: Gascuel, O., Moret, B.M.E. (eds.) WABI 2001. LNCS, vol. 2149, pp. 189–203. Springer, Heidelberg (2001)
26. Siepel, A.: An algorithm to enumerate sorting reversals for signed permutations. J. Comput. Biol. 10, 575–597 (2003)

Inferring a Duplication, Speciation and Loss History from a Gene Tree
(Extended Abstract)[*]

Cedric Chauve[1,2], Jean-Philippe Doyon[3], and Nadia El-Mabrouk[3]

[1] Department of Mathematics, Simon Fraser University, 8888 University Drive,
V5A 1S6, Burnaby (BC), Canada
`cedric.chauve@sfu.ca`
[2] CGL and LaCIM, UQAM, Montréal, Canada
[3] DIRO, Université de Montréal, CP6128, succ. Centre-Ville, H3C 3J7, Montréal
(QC),Canada
`{mabrouk,doyonjea}@iro.umontreal.ca`

Abstract. We consider two questions related to the evolution of gene families. First, given a gene tree for a gene family, can the evolutionary history of this family be explained with only speciation and duplication events, and without gene loss. We show that this question can be answered in linear time, and that such a gene tree induces a single species tree consistent with a history with no loss. We then present a heuristic for the following problem: if a gene tree can not be explained without gene loss, what is the minimum number of losses involved in an evolutionary history of the gene family. We finally evaluate our algorithms on a dataset of plants gene families.

1 Introduction

The duplication of genetic material, from a single gene to the whole-genome, is a fundamental process in the evolution of species, and in particular eukaryotes [12,6]. As a consequence, in most nuclear genomes, many genes are present in multiple copies, that define *gene families*. Gene families evolve, from a single ancestral gene, through microevolutionary events at the nucleotide level, and macroevolutionary events at the genomic level, such as gene duplication, gene loss, genome rearrangements, and speciation events (see [5] and references there). Understanding the evolution of gene families is a fundamental problem that has several applications. For example, it can help to distinguish between orthologs and paralogs: orthologs are copies that are directly related through speciation, while paralogs are copies that have evolved by duplication following a speciation event. This is an important question for functional annotation of genes, as it is believed that pairs of orthologs are more likely to have similar functions. For whole genome analysis based on gene orders and rearrangements, understanding the evolution of gene

---

[*] Work supported by grants from NSERC and SFU.

G. Tesler and D. Durand (Eds.): RECOMB-CG 2007, LNBI 4751, pp. 45–57, 2007.
© Springer-Verlag Berlin Heidelberg 2007

families can help establishing unambiguous one-to-one mappings between pairs of genomes, which is, in general, a hard computational problem (see [2]).

As the notion of orthology and paralogy is directly related to the history of speciation and duplication events during genomes evolution, a natural way of distinguishing between the two types of gene homologues is to infer these events from the phylogenetic tree of a gene family. This question has been widely considered in the case of a well established species tree. It can be described as "fitting a gene tree into a species tree", which is not obvious due to the possible incongruence between the two trees [10]. The main algorithmic approach developed to solve this problem, the gene tree/species tree *reconciliation*, allows to identify the duplications with respect to the speciation events in the species tree [4,13]. It is based on a mapping of the gene tree into the species tree, that can be done in linear time [15,3,16].

Here we consider the more general case where the species tree is unknown. In this context, a natural question is to infer a species tree from a set of gene trees, that optimizes a given criterion, either combinatorial, like the number of duplications and/or losses [13,11], or probabilistic [1]. However, we follow a different approach, as we start from a decision question, motivated among other reasons, by the importance of duplications and speciations to infer co-orthologs: given the gene tree of a specific gene family, can this gene tree be explained using only duplication and speciation events (e.g. without gene losses)? If a gene tree can be explained by a Duplication/Speciation history, we call it a DS-tree (a terminology inspired from [9]). Otherwise, we explain the non-agreement between the gene tree and any DS history by the presence of gene loss events, and we consider the problem of minimizing such number of gene losses. The more general problem of minimizing duplications and losses (in the minimum mutation cost model) for reconciling a set of gene trees has been shown to be NP-hard [13].

In Section 2, we define the notion of a DS-tree in both the frameworks of evolution and reconciliation. We then show, in Section 3, that deciding if a gene tree is a DS-tree can be answered in linear time[1], and that in such a case, there is a single species tree that is compatible with the corresponding Duplication/Speciation history. In the case where a given gene tree T is not a DS-tree, T can be derived from a DS-tree by a series of gene losses. We introduce, in Section 4, the problem of finding the minimum number of gene losses that are needed to transform a DS-tree into T, and we give an efficient heuristic for this problem running in time $O(g \times n)$, where n is the size of T and g is the number of genomes represented in T. We finally analyze, in Section 5, a dataset of plant gene families taken from [14].

2   Duplication/Speciation History and Reconciliation

Duplication/Speciation history. Let $\mathcal{G} = \{1, 2, \cdots, g\}$ be a set of integers representing g different species (genomes). A species tree for $\mathcal{G}$ is a binary tree with exactly g labeled leaves, where each $i \in \mathcal{G}$ represents the label of a single leaf. A gene tree T on $\mathcal{G}$ is a binary tree with labeled leaves, where each leaf is labeled

---

[1] For space reasons, all proofs are omitted and will appear in the full version of this paper.

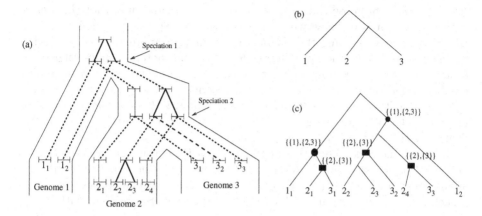

Fig. 1. (a) A DS-history; the segments represent the individual genes; the duplication events are indicated by bold lines, and the speciation events by dashed lines; the genes are denoted as k_i meaning "gene i in genome k". (b) The induced species tree S. (c) The induced gene tree T; notations introduced in Section 3: the partition associated to each internal node is shown; the border B of T contains the two nodes indicated by plain circles, with the associated partition $\{\{1\}, \{2, 3\}\}$, the nodes indicated by plain squares form the border of the forest F_r containing the subtrees of T whose leaves belong to $\{2, 3\}$.

by an integer from $\mathcal{G}$. It is a formal representation of a phylogenetic tree of a gene family, where each leaf labeled i represents a member of the gene family located on genome i.

We say that T is a *Duplication/Speciation tree* (or simply a DS-tree) if there exists a history involving only duplication and speciation events that can lead to the observed tree T. Hereafter, we formally define a Duplication/Speciation history (from now called DS history). See Figure 1 for an illustration.

Definition 1. Let $\mathcal{T} = (T^1, T^2, \cdots T^n)$ be an ordered sequence of n gene trees. We denote by g_k the number of genomes represented by T^k for any k. We say that $\mathcal{T}$ is a DS history if and only if:

1. $T^1 = x$ is a tree restricted to a single vertex x and $g_1 = 1$;
2. For $0 < k < n$, one of the two following situations hold:
   (a) *Duplication event:* T^{k+1} is obtained from T^k by adding two children y and z to a leaf x, and labeling them as x.
   (b) *Speciation event:* There exists i, $1 \leq i \leq g_k$, such that T^{k+1} is obtained from T^k by adding two children y and z to each leaf x of T^k labeled i, and labeling one of the two new nodes by i and the other by $g_k + 1$. Moreover, $g_{k+1} = g_k + 1$.

Let $\mathcal{T}$ be a DS-history leading to a gene tree T. Then, by construction, $\mathcal{T}$ leads to a unique species tree S induced by the speciation events (see Figure 1.a. and b.). We say that the species tree S is *DS-consistent with T*.

Reconciliation. Suppose that a species tree S is already known for $\mathcal{G}$. Then a natural question is to know whether S is DS-consistent with T. This question can be answered by using the classical reconciliation approach that "embeds" the gene tree T into the species tree S [7,13]. The potential non-congruence between a species tree and a gene tree can then be explained by a minimum number of gene losses. More precisely, the reconciliation approach aims to infer a duplication/loss history that has led to the gene tree T, based on a particular mapping (the LCA mapping) from the vertices of T to the vertices of S. We denote by $\ell(T, S)$ the number of loss events.

In this framework, the notion of DS-tree and DS-consistent species tree can be stated as follows (see [7] for a proof of the equivalence between the two approaches): a gene tree T on $\mathcal{G}$ is a DS-tree if there exists a species tree S on $\mathcal{G}$ such that $\ell(T, S) = 0$, in which case S is said to be DS-consistent with T.

3   Recognizing a DS-Tree

In this section, we propose two characterizations of a DS-tree following from the fact that a DS-tree should lead to a species tree S that is DS-consistent with T. The first follows a bottom-up approach, and is the base of the linear-time recognition algorithm presented at the end of this section. The second characterization follows a top-down strategy and leads naturally to our heuristic for the problem of inferring the minimum number of gene losses required to recover a DS-tree from a given gene tree (Section 4).

We first introduce a few notations and definitions. Let T be a gene tree on a genome set $\mathcal{G} = \{1, \ldots, g\}$. For a given vertex x of T, we denote by T_x the subtree of T rooted at x, and by $L(x)$ the subset of $\mathcal{G}$ defined by the labels of the leaves of T_x. We also denote by x_l and x_r respectively the left and right child of x.

A *cherry* of T is a subset $\{i, j\}$ of $\mathcal{G}$ such that $L(x) = \{i, j\}$ for a given vertex x of T.

Definition 2. A cherry $\{i, j\}$ is said to be a *DS-valid cherry* for T if, for any vertex x_l such that $L(x_l) = \{i\}$ (resp. $\{j\}$) and $L(x) \neq \{i\}$ (resp. $\{j\}$) where x is the parent of x_l, the sibling x_r of x_l is such that $L(x_r) = \{j\}$ (resp. $\{i\}$).

If $\{i, j\}$ is a DS-valid cherry, we denote by $c(T, i, j)$ the gene tree on $\mathcal{G} \backslash \{i, j\} \cup \{g + 1\}$ obtained by replacing every internal vertex x such that $L(x) = \{i, j\}$ by a leaf labeled $g + 1$.

Let x be an internal vertex of T. The unordered pair $\{L(x_l), L(x_r)\}$ is called the *partition associated to* x. We say that x *is valid* iff $L(x_l) \cap L(x_r) = \emptyset$. Let F be a forest, that is a set of one or more trees. We say that a set X of vertices of F is *covering* F iff each leaf belonging to a tree of F is a descendant of a unique vertex of the set X. We say that a vertex x is *higher* than a vertex z if z is a descendant of x. Let $B = \{b_1, \ldots, b_k\}$ be the set of highest valid vertices of a forest F: B is called a *border* iff it is covering F and all the partitions associated to the vertices of B are identical. Let B be a border of a forest F, and $\{P_l, P_r\}$ be the partition generated by the vertices of B. We denote by F_l (resp. F_r) the set of subtrees whose leaves labels belong to P_l (resp. P_r) (see Figure 1.c. for an illustration of notations).

Definition 3. A *DS-valid forest* is recursively defined as follows:

1. It is a set of leaves or
2. It has a border and its resulting forests F_l and F_r are DS-valid.

Theorem 1. *Let T be a gene tree on $\mathcal{G}$. The following statements are equivalent.*

1. *T is a DS-tree.*
2. *Either $g = 1$, or for any cherry $\{i, j\}$, $\{i, j\}$ is a DS-valid cherry for T and $c(T, i, j)$ is a DS-tree on $\mathcal{G} \backslash \{i, j\} \cup \{g + 1\}$.*
3. *T is a DS-valid forest.*

Corollary 1. *Let T be a DS-tree on $\mathcal{G}$. There exists a single species tree for $\mathcal{G}$ that is DS-consistent with T.*

Point 2 of Theorem 1 immediately translates into a simple algorithmic principle allowing to check whether a gene tree is a DS-tree. It is based on iteratively considering a cherry, checking its DS-validity, and then contracting all its occurrences into leaves and updating the species tree with the current cherry. We describe below a linear time and space algorithm based on this principle, taking as input a gene tree T on $\mathcal{G}$ with $|\mathcal{G}| = g$, and returning the species tree that is DS-consistent with T, if any.

Algorithm DS-recognition (T)
1.   Let S be an empty tree and $m = g + 1$
2.   Perform a depth-first traversal of T, and let x be the current vertex
3.   IF x is an internal vertex with children x_l and x_r such that
4. $L(x_l) = \{i\}$ and $L(x_r) = \{j\}$ and $i \neq j$ THEN
5.   FOR EVERY vertex z_l such that $L(z_l) = \{i\}$ DO
6.   Let z_r be the sibling of z_l and z its parent
7.   IF $L(z_r) = \{j\}$ THEN replace T_z by a leaf labeled m
8.   ELSE IF $L(z_r) \neq \{i\}$ THEN RETURN FALSE
9.   IF there remains a vertex x with $L(x) = \{j\}$ THEN
10.   RETURN FALSE
11.   Add to S a subtree with root labeled m and children labeled i and j
12.   Increment m
13. RETURN S

Theorem 2. *Given a gene tree T with n vertices, Algorithm DS-recognition returns FALSE iff T is not a DS-tree, and the only species tree that is DS-consistent with T otherwise. It can be implemented to run in $O(n)$ time and space.*

4   Inferring Gene Losses in a Non DS-Tree

4.1   Problem Statements

If a gene tree T is not a DS-tree, and assuming that the given gene tree T is correct (see [8] for a discussion on the case where gene duplications can lead to an incorrect gene tree for a gene family), this implies that some homologous genes are missing or have been deleted or transformed to pseudo-genes during evolution. When a species tree S is known, the reconciliation method can be used to infer a scenario of minimum number $\ell(T, S)$ of gene losses that has led to the observed tree. In this section, we assume that the species tree is unknown, and consider the following natural optimization problem.

Duplication/Loss problem: Given a gene tree T that is not a DS-tree, find a species tree S such that $\ell(T, S)$ is minimum.

This problem can be related to those considered in [13] that compute, for a given gene tree T (or more generally a set of gene trees $T_1, \ldots, T_k$), a species tree S minimizing the total number of duplications (in the so-called duplication cost model) or duplications and losses (in the mutation cost model). They have both been shown to be NP-hard [13], but fixed-parameter tractable [11].

We will instead consider an equivalent formulation of this problem, based on the following property: if T is not a DS-tree, then for every species tree S, there is a DS-tree T^S that can be obtained from T by inserting a minimum number of subtrees such that S is DS-consistent with T^S. Each of these *subtree insertions* represents a gene loss in a given ancestral or extent genome. This way to relate T to a DS-tree T^S, for a given species tree S, leads to the following optimization problem, in the case of an unknown species tree:

Subtrees Insertion Problem: Given a gene tree T that is not a DS-tree, find the minimal number δ of subtree insertions in T allowing to transform T into a DS-tree T'. We denote by (S, δ) a solution to this problem, where S is such that $T' = T^S$.

It follows from [7] that:

Proposition 1. *The Duplication/Loss Problem and the Subtrees Insertion Problem are equivalent: a species tree S is a solution to the Duplication/Loss Problem with $\ell(T, S) = \delta$ if and only if (S, δ) is a solution to the Subtrees Insertion Problem.*

4.2   A Heuristic for the Subtrees Insertion Problem

We now describe an algorithm allowing to obtain an upper bound on the minimum number of subtrees insertions – called *insertions* from now for short – required to transform a gene tree T into a DS-tree.

The method can be decomposed in three steps: (1) recursively label the vertices of T with subsets of the genome set $\mathcal{G}$, (2) use these labels to construct a DS-tree from T, and (3) factorize some of the insertions to reduce the total number of insertions.

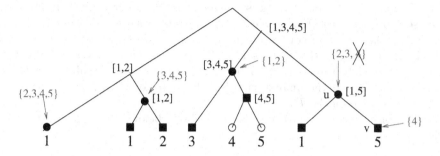

Fig. 2. An illustration of *Procedure Relabel* for a tree on the genome set $\mathcal{G} = \{1, 2, 3, 4, 5\}$. For each internal vertex x, the label in square brackets is the genome set $L(x)$ of x and the label in brackets is the genome subset inserted by Procedure Relabel. This tree has three levels: the first level is the set of vertices indicated by bold circles, the second is the set of bold square vertices, and the third is the two leaves indicated by white circles. The crossed genome in the label of vertex u is the genome removed after applying instruction 7 of *Procedure Relabel*.

Labeling the vertices of T. Initially, each vertex x of T is labeled by its genome set $L(x)$. A set of vertices is said *consistent* if and only if any two vertex labels with a non-empty intersection are identical. *Procedure Relabel* below then relabels the vertices of T in order to obtain successive levels of consistent vertices. It uses the following concepts: a set $\{x_1, \ldots, x_k\}$ of vertices is said to be *connected* if the intersection graph induced by the labels of these vertices (the nodes of the graph are the x_i's and two nodes are connected if their labels have a non-empty intersection) is connected. *Completing* the labels of a set $\{x_1, \ldots, x_k\}$ of connected vertices consists in adding to the label of every vertex x the subset $\cup_{i=1}^{k} L(x_i) \backslash L(x)$ of $\mathcal{G}$, in order that its new label is $\cup_{i=1}^{k} L(x_i)$ (see Fig. 2).

Procedure Relabel (T)
1. F is the forest restricted to the tree T;
2.   WHILE F is not restricted to a set of leaves DO
3.   Let $\mathcal{V}$ be the set of highest valid vertices of F;
4.   Complete the labels of every maximal connected subset of $\mathcal{V}$
5.   Let F be the forest of $\mathcal{V}$;
6.   IF g is inserted in the labels of a vertex x and of a descendant of x THEN
7.   Remove g from the label of x.

The successive sets of highest valid vertices of T considered in *Procedure Relabel* are called the successive *levels of T*. An illustration of this procedure is given in Figure 2.

A first transformation of T into a DS-tree. We now describe how to use the vertices labels computed by *Procedure Relabel (T)* in order to insert subtrees into T, in such a way that the result is a DS-tree. We denote by L the new labeling of the vertices of T computed by *Procedure Relabel (T)*: for a vertex

x of T, $L(x)$ is the new genome set associated to x. For a given level of T, represented by a forest $\mathcal{F}$ of p trees $T_1, \ldots, T_p$ rooted at the vertices $x_1, \ldots, x_p$, we extend the notion of connected subset of the $x_i's$ used in *Procedure Relabel* as follows: a subset of the $T_i's$ is said to be connected if the intersection graph induced by the labels of the corresponding $x_i's$ is connected. We call the *partition of $\mathcal{F}$ by genome sets* the unique partition of $\mathcal{F}$ into forests defined as maximal connected subsets of the $T_i's$.

The construction algorithm is described below, and illustrated in Figure 3.

Procedure Construct-DSTree (T,L)
1.   FOR each level of T (involving insertions) beginning with the last level DO
2.   Let $\mathcal{F}$ be the forest representing the current level;
3.   Let $F_1, \cdots F_p$ be the partition of $\mathcal{F}$ by genome sets;
4.   FOR each subforest F_i DO
5.   Let $\mathcal{G}_i$ be the genome set of F_i;
6.   Let P be an arbitrary phylogeny for $\mathcal{G}_i$;
7.   FOR each vertex x of T such that $L(x) \subset \mathcal{G}_i$ DO
8.   Perform the unique set of subtrees insertions,
9.   in the subtree T_x and on the edge from x to its parent
10.   leading to the phylogeny P

The construction procedure can be reformulated as follows: consider successively each level $\mathcal{F}$ of T beginning with the last one, for each tree T_i of $\mathcal{F}$ rooted at x_i, perform the subtrees insertions leading to the genome set $L(x_i)$, and then replace each tree of T_i by a single leaf. The key observation is that each level considered by this procedure consists solely of leaves and cherries. Therefore, for any x_i, any arbitrary phylogeny P representing the genome set $L(x_i)$ can be obtained by subtrees insertions in T_i. In other words, at each level, there is a coherent way of inserting missing genes in a way leading to the same phylogeny at each node of the tree. This is the main argument used in the proof of the following theorem.

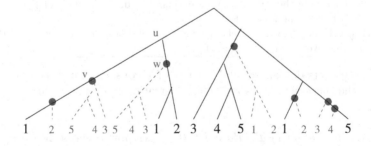

Fig. 3. The result of applying *Procedure Construct-DSTree* on the input (T, L) given by Fig 2. The inserted branches are indicated by dotted lines.

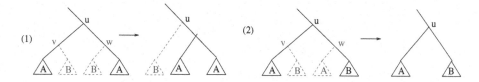

Fig. 4. Illustration of the two factorization rules

Theorem 3. *Let T be a gene tree that is not a DS-tree, and (T, L) the output of* Procedure Relabel(T). *The gene tree computed by* Procedure Construct-DSTree(T,L) *is a DS-tree.*

Corollary 2. *The number of insertions performed by* Procedure Construct-DSTree *is an upper bound on the minimum number of insertions necessary to transform T into a DS-tree.*

Note that in step 6 of *Procedure Construct-DSTree*, we choose an arbitrary phylogeny for the considered subset of taxa. This point could be improved as this phylogeny can be non optimal in terms of the number of subtrees insertions. It could be approached, for example, in a greedy way by selecting the phylogeny that induces the minimum number of subtrees insertions.

Reducing the number of subtrees insertions. A further improved upper bound can be obtained by "factorizing" the subtree insertions made by *Procedure Construct-DSTree*. Let T' be the tree computed by *Procedure Construct-DSTree*, u be an internal vertex of T' that is also a vertex of T. Let $L(u_l)$ be the left genome set of u in T and $L(u_r)$ be the right genome set of u in T. Suppose that the two children v and w of u in S are two inserted vertices, and let $(v, L(v_r))$ and $(w, L(w_l))$ be the two inserted branches. Then we perform the following modification of T':

1. If $L_u = R_u$, then remove the two branches $(v, L(v_r))$ and $(w, L(w_l))$ and insert the subtree T_{w_l} on the branch from u to its parent (see (1) of Figure 4).
2. If $L(v_r) = L(w_l)$ and $L(v_l) = L(w_r)$, then remove the vertices v and w and the subtrees T'_{v_r} and T'_{w_l} (see (2) of Figure 4).

Applying the factorization rule (1) on the tree of Figure 3 gives rise to the tree of Figure 5, leading to 6 subtrees insertions.

Complexity. Let n be the size (number of vertices) of T. A depth-first traversal of T, in time $O(n)$, is required before applying *Procedure Relabel* for the initial labeling of T's vertices by their genome sets. Finding the sets of highest valid vertices then requires a second preorder tree traversal, and for each set of highest valid vertices, relabeling the vertices requires to compare their genome sets, which can be done in time proportional to the number g of different genomes. Therefore, *Procedure Relabel* can be done in time $O(g \times n)$.

For each level $\mathcal{F}$ of T, *Procedure Consruct-DSTree* requires to partition $\mathcal{F}$ into its subforest, which is done in time proportional to g by comparing genome

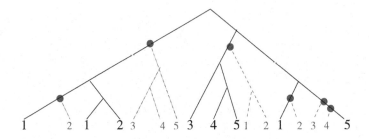

Fig. 5. The result of applying the factorization rules on the DS-tree of Figure 3

sets for one level, and thus in time $O(g \times n)$ for all the levels of T. On the other hand, as each tree insertion can be done in constant time, and a maximum of g tree insertions are performed at each vertex of T, the time complexity for tree insertions is in $O(g \times n)$. Therefore, *Procedure Consruct-DSTree* can be done in time $O(g \times n)$. Finally, the step of reducing the number of subtree insertions can be done in time $O(n)$, by performing a depth-first traversal of T. Therefore, the time complexity of the whole algorithm is in $O(g \times n)$.

5   Experimental Results

We describe the results obtained with the algorithms presented in the previous sections on the 577 gene families studied in [14], in a study of the phylogeny of seven angiosperm genomes from EST data. We focus mainly, in this preliminary experiment, on the computational properties of our algorithms, and in particular on the quality of our heuristic for the Duplication/Loss Problem.

Data. Each of the 577 gene families contains at least four genes and spans at least three genomes. The gene trees were obtained with PAUP, using a maximum likelihood approach (see [14] for a detailed description of the process followed to obtain these gene families and gene trees). The data, including the gene trees and statistics on the size of 577 gene families, and the results of our experiments are available on a companion website, accessible at http//www.lacim.uqam.ca/~chauve/CG07.

Results. First, we found that 333 of the 577 gene trees are DS-trees. However, without surprise, most of these families exhibit few gene duplications. For example, 89 of these 333 families contain 4 genes and span 3 species, while only 7 of the 59 gene trees that span the 7 species are DS-trees (see the file **STATS.txt** on the companion website for the complete statistics). Next, we applied to the 244 remaining gene trees (called non-DS trees from now) the heuristic described in Section 4 to compute an upper bound on the minimum number of gene losses needed to explain the observed gene tree. The results are summarized in the left graphics of Figure 6, and show that many gene families can be explained with few gene losses. We also implemented a branch-and-bound algorithm (to be

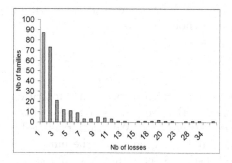 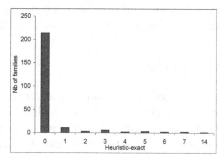

Fig. 6. Left: Distribution of the 244 gene families having a non-DS tree (y axis) according to the number of gene losses inferred by our heuristic (x axis); Right: Distribution of the 244 gene families having a non-DS-tree (y axis) according to the difference between the minimum number of losses needed to transform them into a DS-tree, and the upper bound obtained by our heuristic (x axis).

described in the full version of this paper) in order to assess, on this particular dataset, the quality of our heuristic. The results, summarized in the right graphics of Figure 6, show that it performs well, as for 214 gene trees, it computed the optimal number of gene losses. Finally, our branch-and-bound algorithm allowed us to compute, for each gene tree that is not a DS-tree, the number of species trees that induce a minimum number of gene losses. As shown in the table below, in most cases, there is a unique optimal species tree.

Discussion. On the considered dataset, we found that many gene families could be explained with few gene losses. However, as these gene families are obtained from EST data, it is very likely that many of them are incomplete and it should be expected that more gene losses are required to explain the true evolution of these families. The distribution of the size of the gene families that can be explained without gene losses illustrates this point, as most of them are quite small, while most gene families with many genes and/or genomes require significantly more gene losses.

From an algorithmic point of view, these experiments suggest that our heuristic works well and that, together with our branch-and-bound algorithm, it gives an efficient way to compute the minimum number of gene losses to explain the evolution of a gene family and the corresponding species tree(s). As a consequence, our approach seems to be an interesting candidate to propose quickly, for a given set of gene trees, a set of species trees (for example the species trees

Table 1. Distribution of the 244 gene families with a non-DS tree (second line) according to the number of species trees inducing a minimum number of gene losses (first line)

Number of species trees:	1	2	3	4	5	6	7	13	15
Number of families:	179	16	34	2	6	3	1	2	1

inferred from gene trees that can be explained with few gene losses) that can be analyzed using tree consensus or supertree methods.

6   Conclusion

We proposed in this paper a study of gene trees with a focus on duplication and speciation events. In particular, we showed that deciding if a gene tree is a DS-tree is not difficult, and lead to a single species tree. We also introduced a new way to study the evolution of a gene family by minimizing the number of gene losses. Our preliminary experimental results suggest that our approach is worth further studies, and should be compared with the two other reconciliation approaches based on minimizing the number of duplications and the number of duplications and losses.

Among the algorithmic open problems that our work suggest, the most natural is the complexity of the Subtrees Insertion Problem. In a different perspective, preliminary results on yeast gene families show that our approach needs to be generalized to non fully resolved gene trees (work in progress). It would also be very useful to consider not only gene losses to complete gene trees, but also tree rearrangement operations that could account for potential errors in the obtained gene trees. Finally, as our approach relies on inferring a (or a set of) species tree(s) for each gene family, it would be useful to measure the significance of such species trees.

References

1. Arvestad, L., Berglung, A.-C., Lagergren, J., Sennblad, B.: Gene tree reconstruction and orthology analysis based on an integrated model for duplications and sequence evolution. In: RECOMB 2004, pp. 326–335 (2004)
2. Blin, G., Chauve, C., Fertin, G., Rizzi, R., Vialette, S.: Comparing genomes with duplications: a computational complexity point of view. IEEE/ACM Trans. on Comput. Biol. and Bioinformatics (to appear, 2007)
3. Chen, K., Durand, D., Farach-Colton, M.: NOTUNG: a program for dating gene duplications and optimizing gene family trees. J. Comput. Biol. 7(3-4), 429–444 (2000)
4. Cotton, J.A., Page, R.D.M.: Going nuclear: gene family evolution and vertebrate phylogeny reconcilied. Proc. R. Soc. Lond. B 269, 1555–1561 (2002)
5. Durand, D., Haldórsson, B.V., Vernot, D.: A hybrid micro-macroevolutionary approach to gene tree reconstruction. J. Comput. Biol. 13(2), 320–3354 (2006)
6. Eichler, E.E., Sankoff, D.: Structural dynamics of eukaryotic chromosome evolution. Science 301(5634), 793–797 (2003)
7. Eulenstein, O., Mirkin, B., Vingron, M.: Comparison of annotating duplication, tree mapping, and copying as methods to compare gene trees with species trees. Mathematical hierarchies and biology, DIMACS Ser. Discrete Math. Theoret. Comput. Sci., Amer. Math. Soc., 37, 71–93 (1997)
8. Fares, M.A., Byrne, K.P., Wolfe, K.H.: Rate asymmetry after genome duplication causes substantial long-branch attraction artifacts in the phylogeny of Saccharomyces species. Mol. Biol. Evol. 23(2), 245–253 (2006)

9. Gorecki, P., Tiutyn, J.: DLS-trees: a model of evolutionary scenarios. Theoretical Comput. Sci. 359(1–3), 378–399 (2006)
10. Guigó, R., Muchnik, I., Smith, T.F.: Reconstruction of ancient phylogenies. Mol. Phylogenet. Evol. 6(2), 189–213 (1996)
11. Hallett, M.T., Lagergren, J.: New algorithms for the duplication-loss model. RE-COMB 2000 , 138–146 (1996)
12. Lynch, M., Conery, J.S.: The evolutionary fate and consequences of duplicate genes. Science 290(5494), 1151–1155 (2000)
13. Ma, B., Li, M., Zhang, L.: From gene trees to species trees. SIAM J. Comput. 30(3), 729–752 (2000)
14. Sanderson, M.J., McMahon, M.M.: Inferring angiosperm phylogeny from EST data with widespread gene duplication. BMC Evol. Biol. 7(Suppl. 1), S3 (2007)
15. Zmasek, C.M., Eddy, S.R.: A simple algorithm to infer gene duplication and speciation events on a gene tree. Bioinformatics 17(9), 821–828 (2001)
16. Zhang, L.: On a Mirkin-Muchnik-Smith conjecture for comparing molecular phylogenies. J. Comput. Biol. 4(2), 177–187 (1997)

How to Achieve an Equivalent Simple Permutation in Linear Time

Simon Gog and Martin Bader

University of Ulm, Institute of Theoretical Computer Science, 89069 Ulm, Germany
simon.gog@uni-ulm.de, martin.bader@uni-ulm.de

Abstract. The problem of *Sorting signed permutations by reversals* is a well studied problem in computational biology. The first polynomial time algorithm was presented by Hannenhalli and Pevzner in 1995 [5]. The algorithm was improved several times, and nowadays the most efficient algorithm has a subquadratic running time [9,8]. *Simple permutations* played an important role in the development of these algorithms. Although the latest result of Tannier et al. [8] does not require simple permutations the preliminary version of their algorithm [9] as well as the first polynomial time algorithm of Hannenhalli and Pevzner [5] use the structure of simple permutations. However, the latter algorithms require a precomputation that transforms a permutation into an *equivalent simple permutation*. To the best of our knowledge, all published algorithms for this transformation have at least a quadratic running time. For further investigations on genome rearrangement problems, the existence of a fast algorithm for the transformation could be crucial. In this paper, we present a linear time algorithm for the transformation.

1 Introduction

The problem of *Sorting signed permutations by reversals* (SBR) is motivated by a genome rearrangement problem in computational biology. The task of the problem is to transform the genome of one species into the genome of another species, containing the same set of genes but in different order. As transformation step, only *reversals* (also called *inversions*) are allowed, where a section of the genome is excised, reversed in orientation, and reinserted. This is motivated by the fact that reversals are the most frequent rearrangement operations in nature, especially for bacterial genomes. The problem can be easily transformed into the mathematical problem of sorting a *signed permutation* (i.e. a permutation of the integers 1 to n, where each element has an additional orientation) into the identity permutation. The elements represent the genes of the genome (or any other kind of marker), whereas the signs indicate the strandedness of the genes. As shorter rearrangement scenarios are biologically more plausible than longer ones, one is interested in a minimum sequence of reversals that transforms one permutation into the identity permutation.

SBR is a well studied problem in computational biology, and the first polynomial time algorithm was presented by Hannenhalli and Pevzner in 1995 [5]. The

G. Tesler and D. Durand (Eds.): RECOMB-CG 2007, LNBI 4751, pp. 58–68, 2007.

algorithm was simplified several times [4,6], and the *reversal distance problem* (in which one is only interested in the number of required reversals) can be solved in linear time [1,3]. In 2004, Tannier and Sagot presented an algorithm for SBR that has subquadratic time complexity [9]. This algorithm first transforms the given permutation π into an *equivalent simple permutation* $\hat{\pi}$ and then calculates a sorting for $\hat{\pi}$. This sorting is subsequently used to sort π. In literature, there are several algorithms for this transformation [5,4], but all of them have at least quadratic time complexity (there is an unpublished linear time algorithm by Tannier and Sagot which uses another technique than our algorithm, personal communication). Although Tannier et al. improved their algorithm such that it does no longer require simple permutations [8], a fast algorithm for the transformation could be crucial for further investigations on genome rearrangements. In this paper, we will provide a linear algorithm for transforming a permutation into an equivalent simple permutation.

2   Preliminaries

A *signed permutation* $\pi = \langle \pi_1, \ldots, \pi_n \rangle$ is a permutation of the integers 1 to n, where each element π is assigned a positive ($\overrightarrow{\pi}$) or negative ($\overleftarrow{\pi}$) orientation. A *reversal* $\rho(i,j)$ reverses the order and flips the orientation of the elements between the i-th and j-th element of the permutation. For example, $\rho(3,5)$ transforms $\pi = \langle \overrightarrow{1}, \overrightarrow{2}, \boxed{\overleftarrow{5}, \overleftarrow{4}, \overleftarrow{3}}, \overrightarrow{6} \rangle$ into $id = \langle \overrightarrow{1}, \overrightarrow{2}, \overrightarrow{3}, \overrightarrow{4}, \overrightarrow{5}, \overrightarrow{6} \rangle$. The latter permutation is called identity permutation of size 6. The problem of sorting by reversals asks for a minimal sequence of reversals $\rho_1, \ldots, \rho_k$ that transforms a signed permutation π into the identity permutation. The length k of a minimal sequence is called the reversal distance $d(\pi)$.

The main tool for the solution of the problem of sorting by reversals is the *reality-desire diagram* (also called *breakpoint graph* [2,7]; see Fig. 1 for an example). The reality-desire diagram $RD(\pi)$ of a permutation $\pi = \langle \pi_1, \ldots, \pi_n \rangle$ can be constructed as follows. First, the elements of π are placed from left to right on a straight line. Second, each element x of π with positive orientation is replaced with the two nodes $2x-1$ and $2x$, while each element x with negative orientation is replaced with $2x$ and $2x-1$. We call these nodes *co-elements* of x where the first is called *left node* of x and the other the *right node* of x. Third, we add a single node labeled with 0 to the left of the left node of the first element and add a single node labeled with $2n+1$ to the right of the right node of the last element. Fourth, *reality edges* are drawn from the right node of π_i to the left node of π_{i+1} ($1 \leq i < n$), from node 0 to the left node of π_1, and from the right node of π_n to node $2n+1$. Fifth, *desire edges* are drawn from node $2i$ to node $2i+1$ ($0 \leq i \leq n$). We can interpret reality edges as the actual neighborhood relations in the permutation, and desire edges as the desired neighborhood relations. The *position* of a node v is its position in the diagram and denoted by $pos(v)$ (i.e. the leftmost node has the position 0, the node to its right has the position 1, and so on). As each node is assigned exactly one reality edge and one desire edge, the reality-desire diagram decomposes into cycles. The number of cycles in $RD(\pi)$

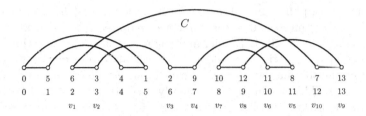

Fig. 1. A reality-desire diagram $RD(\pi)$ for $\pi = \langle \overrightarrow{3}, \overrightarrow{2}, \overrightarrow{1}, \overrightarrow{5}, \overleftarrow{6}, \overleftarrow{4} \rangle$. The first row of numbers are the labels of the nodes, the second are the positions. The third row contains the labeling of nodes of the long cycle C.

is denoted by $c(\pi)$. The *length* ℓ_j of a cycle C_j is the number of desire edges. If ℓ_j is smaller than 3 we call C_j a *short cycle*, otherwise a *long cycle*.

We label the nodes of a cycle C_j as follows. The leftmost node is called $v[j]_1$, then we follow the reality edge to node $v[j]_2$, then follow the desire desire edge to node $v[j]_3$, and so on. We label the reality edge from node $n[j]_{2i-1}$ to $n[j]_{2i}$ with $b[j]_i$ $(1 \le i \le \ell_j)$ and the desire edge from node $n[j]_{2i}$ to $n[j]_{(2i+1)}$ with $g[j]_i$ $(1 \le i < \ell_j)$. The desire edge from node $n[j]_{2\ell}$ to $n[j]_1$ is labeled with $g[j]_{\ell_j}$. If the cycle index j of C_j is clear from the context we omit it.

A desire edge $g = (v_1, v_2)$ is called *oriented* if the positions of v_1 and v_2 in the diagram are both even or odd, otherwise we call g *unoriented*. A cycle which contains no oriented edges is called *unoriented*, otherwise *oriented*.

Two desire edges (v_1, v_2) and (w_1, w_2) *interleave* if the endpoints of the intervals $I_v = [pos(v_1), pos(v_2)]$ and $I_w = [pos(w_1), pos(w_2)]$ are alternating. Two cycles C_1 and C_2 are *interleaving* if there exist interleaving desire edges $f \in C_1$ and $g \in C_2$. A maximal set of interleaving cycles in $RD(\pi)$ is called a *component*. A component is *unoriented* if it contains no oriented cycles, otherwise it is *oriented*.

Hannenhalli and Pevzner found some special structures that depend on unoriented components called *hurdles* and *fortress*. The distance formula for the reversal distance is

$$d(\pi) = n + 1 - c(\pi) + h(\pi) + f(\pi)$$

where $h(\pi)$ is the number of hurdles in $RD(\pi)$ and $f(\pi)$ the indicator variable for a fortress (for details see [5]).

The original Hannenhalli-Pevzner algorithm [5] as well as the subquadratic algorithm of Tannier and Sagot [9] require a permutation whose reality-desire diagram contains only short cycles. Such a permutation is called a *simple permutation*. Hannenhalli and Pevzner showed that every permutation π can be transformed into an *equivalent simple permutation* $\hat{\pi}$, i.e. a simple permutation with $d(\hat{\pi}) = d(\pi)$, by padding additional elements to π. Moreover, a sorting sequence of $\hat{\pi}$ can be used to obtain a sorting sequence of π by ignoring the padded elements.

3   Creating Equivalent Simple Permutations Revisited

We first focus on the creation of simple permutations before we discuss the creation of equivalent simple permutations. If a permutation $\pi = \pi(0)$ has a long cycle, Hannenhalli and Pevzner [5] transform it into a new permutation $\pi(1)$ by ,"breaking" this cycle into two smaller ones. This step is repeated until a simple permutation $\pi(k)$ is achieved.

On the reality-desire diagram the ,"breaking of a cycle" can be described as follows. Let $b = (v_{b1}, v_{b2})$ be a reality edge and $g = (v_{g1}, v_{g2})$ a desire edge belonging to a cycle $C = (\ldots, v_{b1}, v_{b2}, \ldots, v_{g1}, v_{g2}, \ldots)$ in $RD(\pi(i))$. A (b, g)-split of $RD(\pi(i))$ produces a new diagram $\hat{RD}(\pi) = RD(\pi(i+1))$ which is obtained from $RD(\pi(i))$ by:

1. removing edges b and g,
2. adding two new vertices x and y,
3. adding two new reality edges (v_{b1}, x) and (y, v_{b2}),
4. adding two new desire edges (v_{g1}, x) and (y, v_{g2}).

Two examples of such splits are illustrated in Fig. 2. As a result of the split the cycles $(\ldots, v_{b1}, x, v_{g1}, \ldots)$ and $(\ldots, v_{b2}, y, v_{g2}, \ldots)$ are created.

The effect of a (b, g)-split on the permutation can be described as follows. x and y are the nodes of a new element which lies between the consecutive elements previously connected by g. That is, we now consider *generalized permutations* which consists of arbitrary distinct reals instead of permutations of integers. Hannenhalli and Pevzner called the effects of a (b, g)-split on the permutation a (b, g)-padding. We will only use the term (b, g)-split as the two concepts are equivalent.

A (b, g)-split is *safe* if b and g are non-incident, and $\pi(i)$ and $\pi(i+1)$ have the same number of hurdles; i.e. $h(\pi(i)) = h(\pi(i+1))$. The first condition assures that we do not produce a 1-cycle and a cycle with the same size as the old cycle. Because a split is acting on a long cycle, the first condition is easy to achieve. The second condition assures that the reversal distances of $\pi(i)$ and $\pi(i+1)$ are equal (note that a split increases both n and c by one, and the fortress indicator cannot be changed without changing the number of hurdles). The following lemma shows that to fulfill the second condition, it is sufficient to ensure that the resulting cycles belong to the same component.

Lemma 1 ([5]). *Let a (b, g)-split break a cycle C in $RD(\pi(i))$ into cycles C_1 and C_2 in $RD(\pi(i+1))$. Then C is oriented if and only if C_1 or C_2 is oriented.*

In other words, if we do not split a component into two components, the orientation of the component is not changed. For the constructive proof of the existence of safe splits we need the following lemma.

Lemma 2 ([5]). *For every desire edge g that does not belong to a 1-cycle, there exists a desire edge f interleaving with g in $RD(\pi)$. If C is a cycle in $RD(\pi)$ and $f \notin C$ then f interleaves with an even number of desire edges in C.*

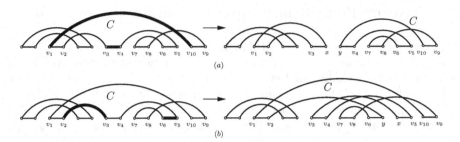

Fig. 2. (a) An unsafe (b,g)-split with $b = (v_3, v_4)$ and $g = (v_1, v_{10})$ that produces a new hurdle. (b) A safe (b,g)-split with $b = (v_5, v_6)$ and $g = (v_2, v_3)$, that does not produce any new components.

And for the linear time algorithm we need the following corollary.

Corollary 1. *Let C be a cycle of length $\ell > 1$ in $RD(\pi)$ with desire edges g_1 to g_ℓ. If these desire edges are pairwise non-interleaving, then there exists a g_j with $1 \leq j < \ell$ and a cycle $C' \neq C$ with a desire edge f, such that f interleaves both g_j and g_ℓ.*

Proof. As C has no pairwise interleaving desire edges, g_ℓ does not interleave with another desire edge of C. So Lemma 2 implies that g_ℓ interleaves with a desire edge f of another cycle C'. Because f is not in C, it interleaves with an even number of desire edges in C. It follows that f interleaves with at least one more desire edge g_j $(1 \leq j < \ell)$ of C.

Theorem 1 ([5]). *If $C = (\ldots, v_1, \ldots, v_{2\ell}, \ldots)$ is a long cycle in $RD(\pi)$, then there exists a safe (b,g)-split acting on C.*

The proof given in [5] is constructive. However, the construction cannot transform the whole permutation into a simple permutation in linear time (which is the goal of our paper). Therefore, in Section 5, we provide an algorithm that achieves this goal in linear time.

4   The Data Structure

We represent the reality-desire diagram as a linked list of $2n + 2$ nodes. The data structure **node** for each node v consists of the three pointers **reality** (pointing to the node connected with v by a reality edge), **desire** (pointing to the node connected with v by a desire edge), and **co_element** (pointing to the co-element of v), and the two variables **position** (the position w.r.t. the leftmost node in the diagram), and **cycle** (the index j of cycle C_j the node belongs to).

We can initialize this data structure for every permutation in linear time. First, the initialization of **reality**, **co_element**, and **position** can be done with a scan through the permutation. Second, for the initialization of **desire** we need the inverse permutation (mapping the nodes ordered by their label to their position) which can also be generated in linear time. Finally, we can initialize cycle by following the reality and desire edges which also takes linear time.

Given a reality edge $b = (v_{b1}, v_{b2})$ and a desire edge $g = (v_{g1}, v_{g2})$, a (b, g)-split can be performed in constant time (see Algorithm 1) if we disregard the problem that we have to update the position variables of the new nodes and all the nodes that lie to the right of b. Fortunately, we need position only to determine if two edges of the same cycle interleave, thus it is sufficient if the relative positions of the nodes of each cycle are correct. This information can be maintained if we set the positions of the new nodes x and y to the positions of the old nodes of b which are now non-incident to x or y. After performing all splits, the reality-desire diagram can easily be transformed into the simple permutation by following desire edges and co-element pointers.

Algorithm 1. (b,g)-split

1: **function** bg-split($b = (v_{b1}, v_{b2}), g = (v_{g1}, v_{g2})$)
2:     create new nodes x, y
3:     $v_{b1}.reality = x$; $v_{b2}.reality = y$ {adjust reality and desire edges}
4:     $x.reality = v_{b1}$; $y.reality = v_{b2}$
5:     $v_{g1}.desire = x$; $v_{g2}.desire = y$
6:     $x.desire = v_{g1}$; $y.desire = v_{g2}$
7:     $x.position = v_{b2}.position$; $y.position = v_{b1}$
8:     $return(x, y)$

---

5   The Algorithm

We now tackle the problem of transforming a permutation into an equivalent simple permutation in linear time. The algorithm has two processing phases.

Phase 1
Our goal in the first phase is to create short cycles or cycles that have no interleaving desire edges. We achieve this goal with a scanline algorithm. The algorithm requires two additional arrays: left[j] stores the leftmost node of each cycle C_j and next[j] stores the right node of the desire edge we are currently checking for interleavings. In both arrays, all variables are initialized with UNDEF. In the following, v_s denotes the current position of the scanline. Before we describe the algorithm, we will first provide an invariant for the scanline.

Invariant: If g_i is a desire edge of the long cycle C_j with $i < \ell_j$, and both nodes of g_i lie to the left of v_s, then g_i does not intersect with any other desire edge of C_j.

It is clear that a cycle C_j has no interleaving edges if the invariant holds and the scanline passed the rightmost node of C_j: g_{ℓ_j} does also not interleave with a desire edge of C_j because the interleaving relation is symmetric. As v_s is initialized with the leftmost node of $RD(\pi)$, the invariant holds in the beginning.

While the scanline has not reached the right end of the diagram, we repeat to analyze the following cases:

Case 1.1 v_s **is part of a short cycle.**
We move the scanline to the left node of the next reality edge. As the invariant only considers long cycles, the invariant is certainly preserved.

Case 1.2 v_s **is part of a long cycle** C_j **and** next[j]=UNDEF.
That is, v_s is the leftmost node of cycle C_j. So we set left[j]=v_s. To check whether $g_1 = (v_2, v_3)$ interleaves with another desire edge, we store the right node of g_1 in next[j] and move v_s to the left node of the next reality edge. Both nodes passed by the scanline (i.e. v_1 and v_2) are the left nodes of a desire edge, so the set of desire edges that lie completely to the left of v_s is not changed and the invariant is preserved.

Case 1.3 v_s **is part of a long cycle** C_j **and** next[j]$\neq v_s$.
Let next[j] be the node v_{2k+1}, i.e. we check for a desire edge that interleaves with g_k (going from node v_{2k} to node v_{2k+1}). As $pos(v_1) < pos(v_{2k}) < pos(v_s) < pos(v_{2k+1})$, there must be a desire edge g_m belonging to C_j that interleaves with g_k. We now distinguish three cases:

(a) g_k **is not** g_1 (for an example, see Fig. 3).
   We perform a (b, g)-split with $b = b_{k+1}$ and $g = g_{k-1}$. That is, we split the 2-cycle $(v_{2k}, v_{2k+1}, x, v_{2k-1})$ from C_j. This split is save since g_k now lies in the 2-cycle that still interleaves with g_m, which belongs to C_j. The right node of the new g_{k-1} in C_j is y, so we adjust next[j] to y.

(b) g_k **is** g_1 **and** g_k **interleaves with** g_{ℓ_j} (see Fig. 4).
   We perform a (b, g)-split with $b = b_1$ and $g = g_2$. That is, we split the 2-cycle (v_2, v_3, v_4, y) from C_j. This split is save since g_1 now lies in the 2-cycle that still interleaves with g_{ℓ_j}, which belongs to C_j. Now, $g_1 = (x, v_5)$, so we set next[j]=v_5. Note that v_5 cannot be to the left of v_s, as v_s is the leftmost node that belongs to C_j and has an index ≥ 4.

(c) g_k **is** g_1 **and** g_k **does not interleave with** g_{ℓ_j} (see Fig. 5).
   It follows that $g_m \neq g_{\ell_j}$. We perform a (b, g)-split with $b = b_2$ and $g = g_{\ell_j}$. That is, we split the 2-cycle (v_2, v_3, x, v_1) from C_j. This split is save since g_1 now lies in the 2-cycle that still interleaves with g_m. As the old leftmost node and reality edge of C_j lie in the 2-cycle we set next[j] = $UNDEF$ which forces the re-initialization of left[j] with v_s and next[j].

In all of these cases, we do not create a desire edge that lies completely to the left of v_s, so the invariant is preserved.

Case 1.4 v_s **is part of a long cycle** C_j **and** next[j]=v_s.
That is, we reach the right node of a desire edge g_k. It follows that g_k does not interleave which any other desire edge of C_j since we have not detected a node of C_j between the left and right node of g_k. Thus moving v_s to the right preserves the invariant. The next desire edge to check is $g_{k+1} = (v_{2(k+1)}, v_{2(k+1)+1})$, so we set next[j] to the right node of g_{k+1} and move v_s to the left node of the next reality edge.

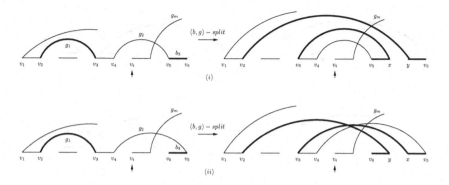

Fig. 3. Case 1.3 (a): (i) $g_k = g_2$ is unoriented or (ii) oriented

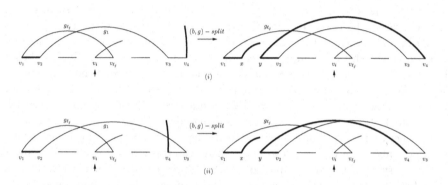

Fig. 4. Case 1.3 (b): (i) g_1 is unoriented or (ii) oriented

We will now analyze the running time of the first phase. In each step we either move the scanline further right (cases 1.1, 1.2, and 1.4) or perform a save (b, g)-split (cases 1.3(a), 1.3(b), and 1.3(c)). As we can perform at most n splits and the resulting diagram can have at most $2n$ reality edges, we have to perform at most $3n$ steps. Each step takes constant time.

Phase 2. After phase 1 we can assure that there remain only short cycles and long cycles with pairwise non-interleaving desire edges. These long cycles have a special structure. The positions of the nodes $v_1, \ldots, v_{2\ell_j}$ of a cycle C_j are strictly increasing and so the first $\ell_j - 1$ desire edges g_i $(i < \ell_j)$ lie one after another. g_{ℓ_j} connects the leftmost and rightmost node of C_j. As we know from Corollary 1 there exists a desire edge f of a cycle $C' \neq C_j$ that interleaves with g_{ℓ_j} and another desire edge g_k of C_j.

We can detect this g_k by first determining a desire edge f which has a node in the interval $I_j = [pos(v_1), pos(v_{2\ell_j})]$ and interleaves with g_{ℓ_j}. Second, we get the g_i that interleaves with f by checking for every desire edge $\neq g_{\ell_j}$ whether it interleaves with f. As I is decomposed by the intervals of the desire edges in distinct areas, we get the corresponding g_i in at most ℓ_j steps.

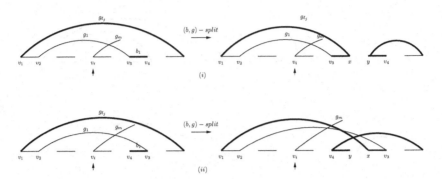

Fig. 5. Case 1.3 (c): (i) g_1 is unoriented or (ii) oriented

Clearly, the second step takes $\sum_{j=1}^{c(\pi)} \ell_j = O(n)$ time. In the first step, we use a stack based algorithm to achieve a linear running time. In each step of the algorithm, the stack will contain a set of intervals I_j of cycles C_j, such that each interval on the stack is completely contained in all other intervals that are below it on the stack (i.e. the topmost interval is contained in all other intervals on the stack). We scan the reality-desire diagram from left to right. For each node v, we check whether its desire edge $f = (v, w)$ interleaves with the topmost interval I_j of the stack. If so, we report the interleaving edges f and g_{ℓ_j}, pop I_j from the stack, check whether f interleaves with the new top interval, and so on, until f does not interleave with the top interval. As the top interval is contained in all other intervals of the stack and Lemma 2 ensures that we find an interleaving edge before we reach the right end of the interval (i.e. v is contained in the topmost interval), f cannot interleave with any other interval on the stack. If v is the leftmost node of a cycle C_j, we push I_j on the stack (note that this interval is equivalent to the desire edge g_{ℓ_j}, so it does not interleave with the topmost interval and is therefore contained in it). In all cases, we continue by moving the scanline one node to the right. The algorithm stops when we have reached the right end of the diagram. During the algorithm, we push the interval I_j of each cycle C_j on the stack, and pop this cycle when we reach a node v in I_j such that the desire edge (v, w) interleaves with I_j. As this node must exist for each cycle (see Lemma 2), we find for each cycle C_j an edge that interleaves with g_{ℓ_j}.

After finding all g_k's we distinguish two cases for a save (b, g)-split:

Case 2.1 $g_{\ell_j - 1} \neq g_k$ (see Fig. 6(i)).
We perform the (b, g)-split on C with $b = (v_1, v_\ell)$ and $g = (v_3, v_4)$. We get $C_1 = (v_1, v_2, v_3, a)$ and $C_2 = (v_\ell, v_{\ell-1}, \ldots, v_4, b)$. As f interleaves with g_1 which is now part of C_1 and g_i which is now part of C_2 the component structure remains the same.

Case 2.2 $g_{\ell_j - 1} = g_k$ (see Fig. 6(ii)).
We perform the (b, g)-split on C with $b = (v_3, v_2)$ and $g = (v_\ell, v_{\ell-1})$. We get $C_1 = (v_1, v_2, b, v_\ell)$ and $C_2 = (a, v_3, v_4, \ldots, v_{\ell-1})$. As f interleaves with g_1 which is now part of C_1 and g_i which is now part of C_2 the component structure remains the same.

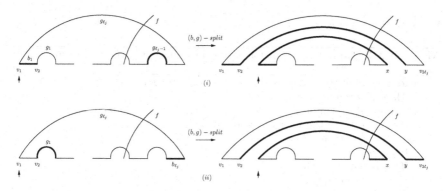

Fig. 6. (i) depicts Case 2.1 and (ii) Case 2.2

In both cases, g_k becomes a desire edge of the cycle C_2, and f intersects both g_k and $g_{\ell'}$ (where ℓ' is the length of C_2). Thus we do not have to recalculate the edge g_k, and can repeat this step on C_2 until the remaining cycles are all 2-cycles. The pseudo code of the whole algorithm is presented in Appendix A. An implementation in $C++$ can be obtained from the authors.

References

1. Bader, D., Moret, B., Yan, M.: A linear-time algorithm for computing inversion distance between signed permutations with an experimental study. Journal of Computational Biology 8, 483–491 (2001)
2. Bafna, V., Pevzner, P.: Genome rearrangements and sorting by reversals. SIAM Journal on Computing 25(2), 272–289 (1996)
3. Bergeron, A., Mixtacki, J., Stoye, J.: Reversal distance without hurdles and fortresses. In: Sahinalp, S.C., Muthukrishnan, S.M., Dogrusoz, U. (eds.) CPM 2004. LNCS, vol. 3109, pp. 388–399. Springer, Heidelberg (2004)
4. Berman, P., Hannenhalli, S.: Fast sorting by reversals. In: Hirschberg, D.S., Meyers, G. (eds.) CPM 1996. LNCS, vol. 1075, pp. 168–185. Springer, Heidelberg (1996)
5. Hannenhalli, S., Pevzner, P.: Transforming cabbage into turnip: polynomial algorithm for sorting signed permutations by reversals. Journal of the ACM 46(1), 1–27 (1999)
6. Kaplan, H., Shamir, R., Tarjan, R.E.: A faster and simpler algorithm for sorting signed permutations by reversals. SIAM Journal on Computing 29(3), 880–892 (1999)
7. Setubal, J., Meidanis, J.: Introduction to Computational Molecular Biology. PWS Publishing Company (1997)
8. Tannier, E., Bergeron, A., Sagot, M.-F.: Advances on sorting by reversals. Discrete Applied Mathematics 155, 881–888 (2007)
9. Tannier, E., Sagot, M.-F.: Sorting by reversals in subquadratic time. In: Sahinalp, S.C., Muthukrishnan, S.M., Dogrusoz, U. (eds.) CPM 2004. LNCS, vol. 3109, pp. 1–13. Springer, Heidelberg (2004)

A   Code

Algorithm 2. Equivalent transformation in a simple permutation into linear time

```
 1: read π and construct the reality-desire diagram RD(π)
 2: mark and count cycles in RD(π)
 3: left[1..c(π)] := {undef,..., undef}; next[1..c(π)] := {undef,..., undef}
 4: set scanline v_s to the leftmost node of RD(π)
 5: while v_s ≠ nil do
 6:     j:=v_s.cycle
 7:     if v_s is part of a short cycle then
 8:         v_s := v_s.reality.co_element
 9:     else if next[j] = undef then {we reach the leftmost point of cycle C_j}
10:         left[j] := v_s
11:         next[j] := v_s.reality.desire
12:         v_s := v_s.reality.co_element
13:     else if v_s = next[j] then {i.e. g_i does not interleave with edge from C_j}
14:         next[j] := v_s.reality.desire
15:         v_s := v_s.reality.co_element
16:     else if g_k is not g_1 then
17:         (x,y):=bg-split(b_{k+1},g_{k-1})
18:         next[j] := y
19:     else if g_k interleaves with g_{ℓ_j} then
20:         (x,y):=bg-split(b_1,g_2)
21:         next[j] := v_5
22:     else {g_k does not interleave with g_{ℓ_j}}
23:         bg-split(b_2,g_{ℓ_j})
24:         next[j] := undef

25: calculate the absolute position for each node in RD(π')
26: create stack ACTIVE_CYCLE
27: set scanline v_s to the leftmost node of RD(π')
28: while v_s ≠ NIL do
29:     while ACTIVE_CYCLE is not empty do
30:         g_ℓ:=ACTIVE_CYCLE.top
31:         if  (v_s, v_s.desire) or (v_s.reality, v_s.reality.desire) interleaves with g_ℓ
             then
32:             determine g_k
33:             ACTIVE_CYCLE.pop
34:         else
35:             break
36:     if v_s is the leftmost node of a long cycle then
37:         ACTIVE_CYCLE.push((v_s, v_s.desire))
38:     v_s := v_s.reality.co_element

39: for each node v_i do
40:     if v_i is the leftmost node of a long cycle C_j then
41:         if g_{ℓ_j-1} ≠ g_k then
42:             bg-split(b_1, g_{ℓ_j-1})
43:         else
44:             bg-split(b_{ℓ_j},g_1)
```

Baculovirus Phylogeny Based on Genome Rearrangements

Daniel Goodman[1], Noah Ollikainen[2], and Chris Sholley[1]

[1] Dept. of Bioengineering, Univerity of California, San Diego
[2] Dept. of Biology, University of California, San Diego

Abstract. Deriving phylogeny of rapidly evolving viral genomes is one of the most challenging problems in evolutionary studies since gene-based phylogenies often produce conflicting evolutionary trees. Phylogenetic reconstruction methods that consider whole genomes are becoming more reliable with the increasing availability of complete genome sequences and the development of algorithms to compare entire genomes. Here we employ a gene rearrangement-based approach to study Baculovirus phylogeny. Since genome rearrangement algorithms require the set of genes shared between all genomes, the most challenging problem in analyzing rapidly evolving genomes is generating this set of genes. Indeed, there are fewer and fewer genes shared between N species as N increases. We address this challenge by iteratively considering smaller sets of related genomes to find conserved genes. Baculovirus was chosen as a test case because a large number of its constituent genomes have been sequenced and its evolutionary relationships are well studied. The resulting phylogenies show clear separation of Baculoviridae into Nucleopolyhedrovirus (NPV) and Granulovirus (GV) as well as the separation of NPVs into groups I and II. Further species separation results in phylogenetic relationships that are largely consistent with conventional gene-based approaches, with some differences that provide insight into the rearrangements of Baculoviridae genomes. Our open source software, *MULGOR* (MULtiple Genome Order), which analyzes genes shared between multiple small rapidly evolving genomes, is available at *http://realm.sdsc.edu/MULGOR/*.

1 Introduction

Traditionally, construction of phylogenetic trees is performed by aligning single genes [15]. However, the usefulness of individual genes in species phylogeny can be limited [24]. Performing sequence alignments of different genes can give conflicting phylogenies. In the case of some rapidly evolving genes, orthologs can be so divergent that their similarities are indistinguishable from randomness [1]. Comparing whole genomes often yields more accurate results for constructing evolutionary trees.

1.1 Phylogenomics and Genome Rearrangement

The availability of whole genome sequences has increased over the past decade, and these data have provided unprecedented insight into eukaryotic, bacterial, archaeal, and viral evolutionary relationships [16]. Recent sequencing of complete genomes has

G. Tesler and D. Durand (Eds.): RECOMB-CG 2007, LNBI 4751, pp. 69–82, 2007.

led to an explosion of phylogenomics: addressing evolutionary problems by considering entire genomes, instead of individual genes [19]. Events that affect the whole genome, as opposed to those that affect single genes, are far less common and thus are more useful for determining phylogeny [24]. These phylogenomic markers have the potential for overcoming many of the problems commonly associated with gene-based phylogenetic analysis.

Analyzing phylogeny by genome rearrangement involves comparing the order of genes in whole genomes to extract phylogeny [2], [3], [8], [10], [21], [22], [26]. An advantage of this method is its reliance on rare evolutionary events with low homoplasy [24].

1.2 Application to Viral Genomes

Previous studies have focused mainly on mammalian genomes [19], leaving viral evolution poorly addressed. Recent studies of viral evolution were based on concatenating all orthologous genes and performing a global sequence alignment [12]. This technique improves accuracy (as compared to single gene analysis) but still does not eliminate all associated problems, especially when dealing with long, distantly related genomes. An alternative technique measures similarity between species by the number of genes that they have in common [27]. This method may produce incorrect phylogenetic relationships in lineages with frequent horizontal gene transfer or rapid gene loss [28].

Hannenhalli et. al., 1995 [10] was the only previous study of viral genomes that used the multiple genome rearrangement method, and was limited to the few herpesvirus genomes available at that time. In the last ten years the number of fully sequenced virus genomes has increased by an order of magnitude, enabling phylogenomic studies of various viruses. About 500 complete double stranded DNA virus genomes are now available from GenBank. The evolutionary volatility of these genomes and their relatively small size (typically less than 400kb) make them good test cases for phylogenomic analysis. We chose baculoviridae as a test family because their evolutionary relationships are well studied [12], [13], [14].

Previous whole-genome studies of Baculovirus phylogeny [12] used a measure called *breakpoint distance* [3] that considers the number of adjacent genes shared between genomes. Here we use the *inversion distance* measure derived from the elegant Hannenhalli-Pevzner theory [9], which considers actual rearrangements in the genome and in so doing correlates more closely with biological events. Phylogenies using inversion distance have been found to be more accurate than breakpoint analysis [18]. Additionally, analysis of inversions suggests evolutionary scenarios at the level of specific rearrangement events and defines a hypothetical ancestral organization [9].

1.3 Finding Shared Genes Presents Key Challenge

When tracking only genome rearrangments, all genomes must be represented in the same alphabet of gene homologs, each gene under consideration must appear once and only once per genome. If a homolog is missing, or present multiple times, it must be removed from all genomes [31]. This has the potential to greatly reduce the number of genes considered, which can in turn reduce the resolution and potentially the accuracy of the phylogeny.

While this problem does not significantly affect the analysis of large genomes (like mammals) it presents a bottleneck for analyzing viruses and other small genomes. Hannenhalli et al. [10] were able to construct their herpesvirus phylogeny without addressing this problem since the number of genomes they studied was small. However, their method does not scale for the 32 Baculovirus genomes examined here: a smaller number of genes are shared between these genomes and thus fewer rearrangements are found.

Multiple genome rearrangement only works when the set of shared genes is not controversial [29]. Indeed, finding sets of orthologous genes conserved between all genomes is one of the key challenges in computing phylogeny by this method [30]. The difficulty is compounded when deletions, horizontal transfer, and duplication of genes are taken into account. Our software, *MULGOR,* automates the discovery of sets of orthologous genes among multiple small genomes, removing the need for manual annotation of these sets or the removal of paralogous groups from consideration.

Sankoff's exemplar method [25] tackles the problem of paralogous genes by removing all but one paralog from each genome such that the reversal distance between these two remaining genes is minimized. This problem was found to be NP-hard [6], although heuristic approaches exist. Because we consider paralogs from more than two genomes at a time, and consider many sets of paralogs simultaneously, our method chooses 'exemplar' genes based on sequence similarity instead of optimized reversal distance.

Some methods that detect orthologous, such as *INPARANOID* [23], work on pairs of genomes only. While *OrthoMCL* [17] is robust and functions on multiple large eukaryotic genomes, we chose to implement a faster and less complex graph-theoretic approach, which works on flat-files and does not require relational databases, to deal with these much smaller viral genomes. It is assumed that our method would not perform as well on larger genomes with more paralogs. If *MULGOR* was to be expanded to support larger genomes, the integration of an algorithm like *OrthoMCL* would likely be ideal.

To deal with a lack of gene conservation (genes present in most but not all genomes), our automated method increases the granularity of gene ordering by considering smaller subgroups of genomes with more shared genes.

1.4 Baculoviridae

The family Baculoviridae has many qualities that make it an interesting and useful test case. The family has a history of horizontal gene transfer and variable rates of mutation between genes that result in discrepancies in single-gene phylogeny [12]. Baculoviruses have speciated on a lengthy evolutionary timeline, and a significant amount of genome rearrangement has occurred between members of the family with different arthropod hosts. Previous baculoviridae studies have employed a variety of methods for phylogenetic reconstruction based on gene content, gene sequence and relative breakpoint distance [12], [13], [14]. This study is the first rearrangement-based analysis on this family. Additionally, over twenty percent of the completely sequenced Baculoviridae genomes were published in the last year alone. This study incorporates these new genomic sequences to study Baculoviridae phylogeny.

2   Materials and Methods

2.1   Detection of Orthologous Gene Clusters

The genome sequences of thirty-two Baculoviruses were obtained from NCBI (Table 1). Prior to determining the relative gene orders of a given set of genomes, genes conserved across all genomes were identified via single-linkage clustering, in a method similar to BLASTCLUST (unpublished, available at http://www.ncbi.nlm.nih.gov/ BLAST/docs/blastclust.html). In this method, each gene was represented as a vertex, and an edge was placed between two vertices if the e-value of their alignment was lower than the cutoff. After all genes were compared, a cluster was defined by the genes representing a connected component in the resulting graph. Clusters that contained at least one gene for every genome considered were kept. While we did not use BLASTCLUST, because it does not save each BLASTP score and does not always perform every pair-wise comparison, our method is basically a reimplementation of the BLASTCLUST algorithm.

Because of duplication and paralogous sets of genes, a second level of homology detection was employed. After initial clustering, larger clusters that contained more than one gene per genome were 'pruned' using a second algorithm described in the following section. An e-value cutoff was chosen manually to minimize cluster overlap and generate the greatest number of ideally sized clusters. For Baculovirus cluster iterations, this e-value was 1e-12. This ideal value varies for different datasets, but tests with Poxviridae and Herpesviridae (data unpublished) suggest that this value works well for many large viral genomes.

2.2   Cluster Pruning

To find rearrangements between some closely-related genomes, it was necessary to include the larger clusters that would otherwise be ignored via the method outlined above. A second algorithm was used to break up and then prune these large clusters. The algorithm models each cluster as a graph where all genes are vertices and all pair-wise BLAST scores are weighted undirected edges (asymmetric scores are averaged).

Clusters in which every genome is present more than twice represent multiple similar sets of orthologs, and these clusters were first broken up via the following method. It was observed that edges on these largest graphs fall into two groups: high weight edges (higher than the mean) within closely related sub-clusters and low weight edges (lower than the mean) between different sub-clusters. Those falling into the latter group were removed from the graph, leaving multiple disconnected subgraphs. The subgraphs that contained at least one gene from every genome were kept.

After breaking up these 'mega-clusters' into subgraphs, a modified version of Prim's algorithm [7] was run to 'prune off' paralogs. We took the highest weighted edges, not the least, and as an extra constraint on expanding the tree, only chose vertices that represented a gene from a genome that was not already present in the tree. The algorithm finished when the tree contained vertices representing one gene from each genome. This was a heuristic approach that nevertheless returned the strongest subsets in all observed cases. In this fashion, multiple usable clusters were extracted from one larger cluster that contained paralogs and multiple ortholgous sets.

Table 1. Baculovirus Accessions

Species	Genome nt	Accession	Date Created
L. dispar nucleopolyhedrovirus (LdMNPV)	161,046	NC_001973	11/3/1998
Xestia c-nigrum granulovirus (XcGV)	178,733	NC_002331	6/7/2000
Trichoplusia ni SNPV virus	134,394	NC_007383	9/7/2005
H. armigera nucleopolyhedrovirus G4 (HaSNPV)	131,403	NC_002654	1/25/2001
Choristoneura fumiferana MNPV	129,593	NC_004778	5/6/2003
Agrotis segetum nucleopolyhedrovirus	147,544	NC_007921	3/27/2006
Agrotis segetum granulovirus	131,680	NC_005839	4/9/2004
Mamestra configurata nucleopolyhedrovirus B	158,482	NC_004117	8/25/2002
Mamestra configurata NPV-A	155,060	NC_003529	3/29/1997
Autographa californica nucleopolyhedrovirus (AcMNPV)	133,894	NC_001623	7/16/1994
Bombyx mori nucleopolyhedrovirus (BmNPV)	128,413	NC_001962	1/18/1996
Cydia pomonella granulovirus (CpGV)	123,500	NC_002816	4/2/2001
Culex nigripalpus Baculovirus (CuniNPV)	108,252	NC_003084	8/22/2001
E. postvittana nucleopolyhedrovirus (EppoNPV)	118,584	NC_003083	8/19/2001
H. zea single nucleocapsid nucleopolyhedrovirus (HzSNPV)	130,869	NC_003349	1/1/2002
O. pseudotsugata multicapsid NPV (OpMNPV)	131,995	NC_001875	3/27/1997
Plutella xylostella granulovirus (PxGV)	100,999	NC_002593	10/29/2000
S. exigua nucleopolyhedrovirus (SeMNPV)	135,611	NC_002169	12/29/1999
Spodoptera litura nucleopolyhedrovirus (SpliNPV)	139,342	NC_003102	9/11/2001
Adoxophyes honmai NPV	113,220	NC_004690	4/5/2003
Adoxophyes orana granulovirus	99,657	NC_005038	7/15/2006
Antheraea pernyi nucleopolyhedrovirus	126,630	NC_008035	5/16/2006
C. fumiferana defective nucleopolyhedrovirus	131,160	NC_005137	10/11/2003
Choristoneura occidentalis granulovirus	104,710	NC_008168	6/19/2006
Chrysodeixis chalcites nucleopolyhedrovirus	149,622	NC_007151	6/29/2005
Cryptophlebia leucotreta granulovirus	110,907	NC_005068	8/13/2003
Helicoverpa armigera nuclear polyhedrosis virus	130,759	NC_003094	8/31/2001
Hyphantria cunea nucleopolyhedrovirus	132,959	NC_007767	2/2/2006
Leucania separata nuclear polyhedrosis virus	168,041	NC_008348	9/16/2006
Neodiprion abietis nucleopolyhedrovirus	84,264	NC_008252	7/24/2006
Neodiprion lecontei NPV	81,755	NC_005906	6/17/2004
Neodiprion sertifer nucleopolyhedrovirus	86,462	NC_005905	6/17/2004
Phthorimaea operculella granulovirus	119,217	NC_004062	7/1/2002
Plutella xylostella multiple nucleopolyhedrovirus	134,417	NC_008349	9/16/2006
Rachiplusia ou multiple nucleopolyhedrovirus	131,526	NC_004323	10/2/2002

2.3   Gene Order and Phylogeny

After obtaining the clusters of one-to-one orthologous genes, the order that these orthologs appear in each genome was determined using the relative starting position and coding strand of each ORF. The gene ordering for each genome was used as input into the MGR (Multiple Genome Rearrangement) program, which generated a phylogenomic tree based on reversal distance [4].

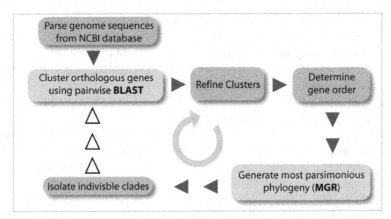

Fig. 1. Diagram of *MULGOR* and how phylogeny is derived from orthologous clusters

2.4   *MULGOR* Software

The above procedures were conducted automatically using *MULGOR* (MULtiple Genome ORder), a set of perl scripts that takes NCBI genome accession numbers as input, downloads the sequences, and generates relative gene orders for the common set of genes from these genomes. *MULGOR* then feeds these gene orders to MGR to produce a tree. The software also includes an option to run pairwise comparisons between genomes and produce 2-dimensional plots of relative gene positions. Genes on the plot can be represented as lines corresponding to sequence length, or as lines of equal length. This feature was used as a preliminary method to gauge the relation between two genomes.

2.5   Using *MULGOR* Iteratively on Baculovirus

First, all genomes for available members of the Baculovirus family were found, and run through the script. Because fewer orthologs were shared by all family members as compared to those shared by individual subgroups, the first pass generated a tree of low resolution, with many genomes sharing the same gene order. These identically-ordered subgroups were then automatically run through the script again, generating more orthologous genes and potentially more rearrangements between them, giving a higher resolution tree. This process was continued iteratively until further iterations could not yield refinements of the tree.

3 Results

3.1 Baculovirus Phylogeny of Nine Species

We first applied our method to derive phylogeny for nine Baculovirus genomes previously studied by Herniou *et al*, 2001 [12]. Our software identified 43 genes common to all nine genomes and constructed the most parsimonious rearrangement scenario, shown as a phylogenetic tree in Figure 2. This tree clearly shows the

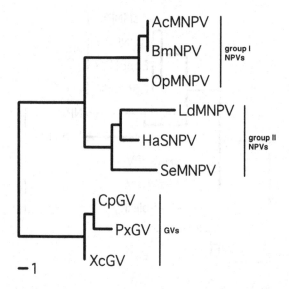

Fig. 2. Rearrangement-based phylogenetic tree of nine Baculovirus genomes, based on 43 orthologous gene clusters. Distances are relative to the number of rearrangement events between genomes. The distance of one rearrangement is shown in the bottom left of the figure.

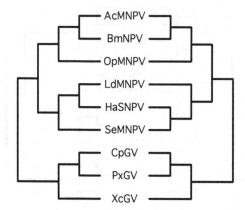

Fig. 3. Comparison of rearrangement-based phylogenetic tree (left) with a tree based on analysis of concatenated gene sequences from Herniou et al., 2001 (right)

separation of Nucleopolyhedrovirus (NPV) and Granulovirus (GV), as well as the splitting of NPV into groups I and II. Figure 3 compares our rearrangement-based tree with a sequence-based tree of the previous study, showing a difference in the phylogenies of group II NPVs.

3.2 Baculovirus Phylogeny of Twelve Species

Our method was further applied to reconstruct the phylogeny of twelve Baculovirus genomes examined by Herniou et al., 2003 [13]. The dipteran NPV, CuniNPV, was

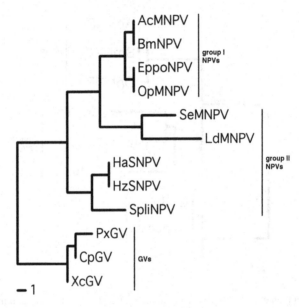

Fig. 4. Rearrangement-based phylogenetic tree of twelve Baculovirus genomes, based on 43 orthologous gene clusters

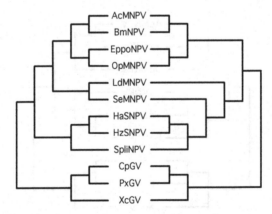

Fig. 5. Comparison of rearrangement-based phylogenetic tree (left) with a tree based on analysis of 30 concatenated gene sequences from Herniou et al., 2003 (right)

excluded from this analysis due to its significant divergence from the other lepidopteran Baculoviruses. A most parsimonious rearrangement tree based on 43 genes shared by the twelve Baculovirus genomes is shown in Figure 4. The division of NPVs and GVs is clear, but the separation of group I and group II NPVs is obscured by an abundance of rearrangements between group II NPVs. A comparison of this tree with the corresponding phylogeny derived with a gene-sequence method used by Herniou, et al, 2003 [13] is shown in Figure 5. While the relationships among group I NPVs and GVs are identical between both trees, there is disagreement in the resolving of group II NPVs.

3.3 Updated Phylogeny of Thirty-two Genomes

To provide an update on Baculovirus phylogeny, we incorporated recently sequenced Baculovirus genomes into our analysis and reconstructed the phylogeny of the resulting 32 Baculovirus genomes. Once again, non-lepidopteran Baculoviruses were excluded in order to maintain a significant number of shared genes. The resulting tree, based on the rearrangements of 35 shared genes, is shown in Figure 6.

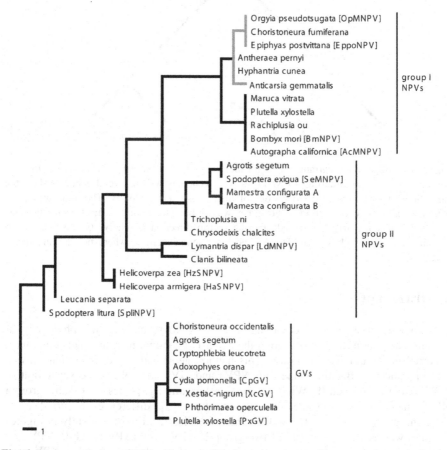

Fig. 6. Phylogenetic tree of 32 lepidopteran Baculovirus genomes. The cyan subtree shows the result of a second iteration necessary to resolve the phylogeny of six group I NPV genomes.

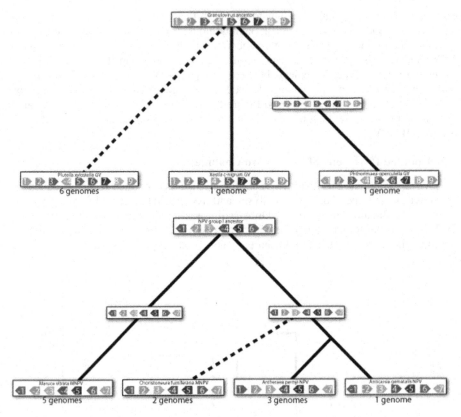

Fig. 7. Tree of genome rearrangements for the three different orderings in *Granulovirus* (top) and the four different orderings of *NPVI* (bottom) genomes. When several genomes share the same order, they are represented by one of their constituent genomes. The total number of genomes for each gene ordering is shown below each box (see Figure 6). Each icon represents one 'strip' of genes with conserved ordering. Strip direction is relative because each strip contains genes on both strands. Dotted lines represent ancestries where no rearrangements occurred.

4 Discussion

A genome-rearrangement method has been used to reconstruct the phylogeny of Baculovirus genomes, demonstrating the extensive genome rearrangements that have occurred in Baculovirus evolution. This method generated phylogenies that showed the separation of Baculoviridae into NPVs and GVs, as well as the separation of NPVs into groups I and II. While the relationships among species of GVs and group I NPVs matched those in previous studies [12], [13], differences occurred when resolving group II NPVs. These differences are illustrated most evidently in Figures 4 and 5, where the gene orders of two group II NPVs, SeMNPV and LdMNPV, are shown requiring fewer rearrangements to be transformed into a group I NPV gene order than into another group II NPV gene order.

4.1 Gene Deletions and Low Sequence Homology

One explanation for the placement of SeMNPV and LdMNPV in Figures 4 and 5 is that the gene orders used did not contain the entire set of genes shared between all Baculoviruses in the tree. The gene orders used consisted of 43 shared genes determined by our software, yet Herniou et. al. 2001, showed the presence of 63 genes shared between these genomes.

The genes that were not identified as shared genes fall into two categories. The first category consists of genes whose similarity to orthologous genes did not exceed our similarity threshold. The second consists of ortholog sets that were missing genes from one or more genomes. In the second case, it is often difficult to tell the if the missing genes were the result of low sequence homology or were simply deleted in those genomes. The larger number of shared genes found in Herniou et. al. 2001 was the result of mining sequence annotations in the literature, which allowed for the identification of distant orthologs that our method currently ignores.

4.2 Deviations from Parsimony

The differences between rearrangement-based and sequence-based trees shown in Figures 3 and 5 could suggest that the most parsimonious rearrangement scenario does not mirror the actual evolutionary events that occurred in Baculovirus evolution. The most parsimonious scenario minimizes the number of rearrangements and is typically assumed to be valid due to low probability of a rearrangement occurring. However, group II NPVs in particular show marked rearrangement, and a scenario with additional rearrangements that results in an altered tree topology is plausible. For instance, in Figure 2, HaSNPV is eight rearrangements from both SeMNPV and LdMNPV, and the most parsimonious scenario shows that HaSNPV and LdMNPV diverged from a common ancestor not shared by SeMNPV. Another scenario with equal or slightly more rearrangements could result in HaSNPV and SeMNPV with a common ancestor not shared by LdMNPV. If MGR were altered to allow for the output of multiple trees that are very close to the most parsimonious, it might facilitate the comparison of sequence- and gene-order based trees to derive the most likely evolutionary scenario.

4.3 Strong Gene Order Conservation vs. Lack of Shared Genes

While a large number of genome rearrangements was identified in group II NPVs, strong gene order conservation was observed among group I NPVs as well as GVs, as seen in Figure 7. Because of the large number of gene 'strips' resulting from the many rearrangements in NPV II, we could not display the rearrangements for this subgroup. The high gene order conservation posed a problem when constructing the 32 Baculovirus genome tree, as some phylogenic relationships could not be determined. Five group I NPVs, including AcMNPV and BmNPV, maintained identical gene orders even when 112 shared genes were considered. Another iteration of the algorithm was performed to separate the group, and is shown in cyan in Figure 6. This was successful to some degree, but three genomes still remained with identical order. This also occurred with the group of five GVs containing CpGV, each having identical orders of 82 shared genes. Such extensive conservation of gene order demonstrates a limitation with rearrangement-based approaches. However, these situations can be overcome by applying gene sequence and gene content methods to groups of genomes with highly conserved order. Tracking deletions and horizontal gene transfers as separate events in genome evolution, alongside rearrangements, might also yield results here.

Just as conservation presents a problem to rearrangement-based methods, lack of conservation also creates challenges. For a gene to be considered in the gene order, it must be present in all genomes being considered. Highly divergent species or species with rapid gene loss cannot be used unless they possess a significant sized core set of genes shared with the other genomes. Non-lepidopteran Baculoviruses were excluded from this study because they greatly reduced the number of shared genes and thus generated low-resolution error-prone phylogenetic trees. Lepidopteran Baculoviruses, however, are similar enough to provide a sufficient number of common genes while being distant enough to trace their evolutionary history through genome rearrangement. Analysis of small genomes with a genome rearrangement-based approach requires a well-balanced compromise between similarity and divergence.

4.5 Improvements and Future Work

To increase the number of identified orthologs, an improved method would use a similarity threshold that varies for each cluster, allowing for distantly related orthologs to be clustered together. In the case of SeMNPV and LdMNPV, these additional genes could have an order similar to other group II NPVs, resulting in a tree identical to the one obtained by Herniou et al. 2004 via gene-sequence methods.

Instead of a top-down multiple iteration approach, a different approach might also be employed. This method would consider smaller sets of genomes that are known to share many conserved genes instead of the set of all genomes, which shares fewer orthologous sets. These smaller triplet phylogenies could then be merged using the gene orders of the triplet ancestor. This bottom-up method might bypass the main difficulty seen here: a lack of gene conservation among disparate genomes.

4.6 Conclusion

While our results are largely consistent with Herniou et. al., there are some important differences, as outlined above. It is not uncommon to have conflicting and controversial scenarios for many questions in evolution, let alone viral evolution. The phylogenies presented in Herniou et. al., 2001, 2003, 2004 also point to self-conflicting evolutionary relationships in some cases. Further, because the phylogenies presented have no bootstrap values or confidence scores, it is difficult to objectively assess their relative accuracy. Aside from the difficulties, our method presents an automated tool that can be used alongside existing gene-content-based methods of phylogeny derivation.

Acknowledgements. The authors would like to thank Dr. Pavel Pevzner for his helpful advice and constructive discussions and the reviewers for their thoughtful and useful comments and suggestions. The work was supported by the Howard Hughes Medical Institute Professor Award.

References

1. Albrecht, J., Nicholas, J., Biller, D., Cameron, K.R., Biesinger, B., Newman, C., Wittmann, S., Craxton, M.A., Coleman, H., Fleckenstein, B., Honess, R.W.: Primary structure of the herpesvirus saimiri genome. J. Virol. 66, 5047–5058 (1992)
2. Belda, E., Moya, A., Silva, F.: Genome Rearrangement Distances and Gene Order Phylogeny in γ -Proteobacteria. Molecular Biology and Evolution. 22, 1456–1467 (2005)
3. Blanchette, M., Bourque, G., Sankoff, D.: Breakpoint phylogenies. In: Miyano, S., Takagi, T. (eds.) Genome Informatics 1997, pp. 25–34. Univ. Academy Press (1997)
4. Bourque, G., Pevzner, P.: Genome-Scale Evolution: Reconstructing Gene Orders in the Ancestral Species. Genome Research. 12(1), 26–36 (2002)
5. Bourque, G., Pevzner, P.A., Tesler, G.: Reconstructing the genomic architecture of ancestral mammals: lessons from human, mouse, and rat genomes. Genome Research. 14, 507–516 (2004)
6. Bryant, D.: The complexity of calculating exemplar distances. In: Sankoff, D., Nadeau, J. (eds.) Comparative Genomics, pp. 207–212. Kluwer Academic Publishers, Dordrecht (2000)
7. Cormen, T.H., Leiserson, C.E., Riverst, R.L, Stein, C.: Introduction to Algorithms. McGraw Hill, New York (2001)
8. El-Mabrouk, N.: Genome rearrangement by reversals and insertions/deletions of contiguous segments. In: Giancarlo, R., Sankoff, D. (eds.) CPM 2000. LNCS, vol. 1848, pp. 222–234. Springer, Heidelberg (2000)
9. Hannenhalli, S., Pevzner, P.: Transforming cabbage into turnip (polynomial algorithm for sorting signed permutations by reversals). In: Theory of Computing STOC 95, pp. 178–189. ACM Press, New York (1995)
10. Hannenhalli, S., Chappey, C., Koonin, E., Pevzner, P.: Genome Sequence Comparison and Scenarios for Gene Rearrangements: A Test Case. Genomics. 30, 299–311 (1995)
11. Hanson, L., Rudis, M., Vasquez-Lee, M., Montgomery, D.: A broadly applicable method to characterize large DNA viruses and adenoviruses based on the DNA polymerase gene. Virology Journal 3, 28 (2006)
12. Herniou, E., Luque, T., Chen, X., Vlak, J., Winstanley, D., Cory, D., O'Reilly, D.: Use of Whole Genome Sequence Data To Infer Baculovirus Phylogeny. Journal of Virology 75(17), 8117–8126 (2001)
13. Herniou, E., lszewski, O., Cory, J., O'Reilly, D.: The Genome Sequence and Evolution of Baculoviruses. Annual Review of Entomology 48, 211–234 (2003)
14. Herniou, E., Olszewski, J., O'Reilly, D., Cory, J.: Ancient Coevolution of Baculoviruses and Their Insect Hosts. Journal of Virology. 78(7), 3244–3251 (2004)
15. Graur, D., Li, W.: Fundamentals of Molecular Evolution. In: Sinauer Associates, Sunderland, Massachusetts (2000)
16. Koonin, E.V., Aravind, L., Kondrashov, A.S.: The impact of comparative genomics on our understanding of evolution. Cell 101(6), 573–576 (2000)
17. Li, L., Stoeckert, C.J., Roos, D.S.: OrthoMCL: Identification of Ortholog Groups for Eukaryotic Genomes. Genome Res. 13(9), 2178–2189 (2003)
18. Moret, B.M.E., Siepel, A.C., Tang, J., Liu, T.: Inversion medians outperform breakpoint medians in phylogeny reconstruction from gene-order data. In: Guigó, R., Gusfield, D. (eds.) WABI 2002. LNCS, vol. 2452, pp. 521–536. Springer, Heidelberg (2002)

19. Murphy, W.J., Larkin, D.M., van der Wind, A.E., Bourque, G., Tesler, G., Auvil, L., Beever, J.E., Chowdhary, B.P., Galibert, F., Gatzke, L., Hitte, C., Meyers, S.N., Milan, D., Ostrander, E.A., Pape, G., Parker, H.G., Raudsepp, T., Rogatcheva, M.B., Schook, L.B., Skow, L.C., Welge, M., Womack, J.E., O'brien, S.J., Pevzner, P.A., Lewin, H.A.: Dynamics of mammalian chromosome evolution inferred from multispecies comparative maps. Science 309(5734), 613–617 (2005)

20. Murphy, W., Pevzner, P., O'Brien, S.: Mammalian phylogenomics comes of age. Trends in Genetics. 20(12), 631–639 (2004)

21. Olmstead, R., Palmer, J.: Chloroplast DNA systematics: a review of methods and data analysis. Amer. J. Bot. 81, 1205–1224 (1994)

22. Palmer, J., Herbon, L.: Plant mitochondrial DNA evolves rapidly in structure, but slowly in sequence. J. Mol. Evol. 27, 87–97 (1988)

23. Remm, D., Storm, C.E.V., Sonnhammer, E.L.L.: Automatic Clustering of Orthologs and In-paralogs from Pairwise Species Comparisons. J. Mol. Biol. 314, 1041–1052

24. Rokas, A., Holland, P.W.H.: Rare genomic changes as a tool for phylogenetics. Trends in Ecology & Evolution 15(11), 454–459 (2000)

25. Sankoff, D.: Genome rearrangement with gene families. Technical Report, Centre de recherches mathématiques, Université de Montréal (1999)

26. Sankoff, D., Blanchette, M.: Multiple genome rearrangement and breakpoint phylogeny. Journal of Computational Biology 5, 555–570 (1998)

27. Snel, B., Bork, P., Huynen, M.: Genome phylogeny based on gene content. Nature Genetics. 21, 108–110 (1999)

28. Snel, B., Bork, P., Huynen, M.: Genomes in Flux: The Evolution of Archaeal and Proteobacterial Gene Content. Nature. 417, 399–403 (2002)

29. Tesler, G., Pavel, P.: Genome Rearrangements in Mammalian Evolution: Lessons From Human and Mouse Genomes. Genome Res. 13, 37–45 (2003)

30. Thornton, J.W., DeSalle, R.: Gene family evolution and homology: genomics meets phylogenetics. Annu. Rev. Genomics Hum. Genet. 1, 41–73 (2000)

31. Wang, L., Warnow, T.: Estimating true evolutionary distances between genomes. In: Proceedings of the Thirty-third Symposium on Theory of Computing (STOC'01), pp. 637–646. ACM Press, New York (2001)

Learning Gene Regulatory Networks via Globally Regularized Risk Minimization

Yuhong Guo and Dale Schuurmans

Department of Computing Science,
University of Alberta, Edmonton T6G 2E8, Canada
{yuhong,dale}@cs.ualberta.ca

Abstract. Learning the structure of a gene regulatory network from time-series gene expression data is a significant challenge. Most approaches proposed in the literature to date attempt to predict the regulators of each target gene individually, but fail to share regulatory information between related genes. In this paper, we propose a new globally regularized risk minimization approach to address this problem. Our approach first clusters genes according to their time-series expression profiles—identifying related groups of genes. Given a clustering, we then develop a simple technique that exploits the assumption that genes with similar expression patterns are likely to be co-regulated by encouraging the genes in the same group to share common regulators. Our experiments on both synthetic and real gene expression data suggest that our new approach is more effective at identifying important transcription factor based regulatory mechanisms than the standard independent approach and a prototype based approach.

1 Introduction

Genes and their products do not work independently in the cell. Rather, they are jointly regulated in a coordinated fashion, both internally and externally, to achieve proper cell function. One of the key mechanisms of gene regulation takes place at the mRNA transcription level. With the emergence of high-throughput microarray techniques, the mRNA expression levels of thousands of genes can be measured simultaneously. Using computational techniques to learn gene regulatory networks from high-throughput time-series gene expression data has been an active area of research in recent years. The goal of such research is to discover the causal control relationships between genes, which would offer a fundamental understanding of how biological processes are coordinated in the cell.

A variety of computational approaches have been proposed in the literature to model gene regulatory networks from expression data. Many approaches have been based on the use of linear models to express dependence between time series profiles. For example, D'Haeseleer et al. [1] studied a straightforward linear model for this purpose; Chen et al. [2] and De Jong et al. [3] investigated linear differential equations for gene regulatory network modeling. All of these approaches suffer from risks of over-fitting, however, since they fit a number of parameters that is proportional to the size of the data itself. To counter the risk

G. Tesler and D. Durand (Eds.): RECOMB-CG 2007, LNBI 4751, pp. 83–95, 2007.
© Springer-Verlag Berlin Heidelberg 2007

of over-fitting, other linear approaches have taken advantage of sparseness of the regulatory relationship between genes; that is, that any one gene is regulated by a small subset of the other genes. De Hoon et al. [4] have proposed to use "Akaike's Information Criterion" (AIC) to determine the nonzero coefficients in the linear system. Similarly, Li & Yang [5] used "L1 regularization" to conduct feature selection on the linear parent set.

Another popular approach to learning gene regulatory network structure is to exploit various forms of (dynamic) Bayesian network structure learning methods. A Bayesian network is a graphical representation of the causal relationships underlying a set of variables that provides a sound probabilistic framework for representing and inferring probabilistic relationships. *Dynamic* Bayesian networks are a natural extension of Bayesian networks to modeling time-series data. Learning the structure of a Bayesian network from data generally requires one of two approaches to be followed: a score-based approach—where a heuristic search is performed through the space of causal network structures to identify the most likely structure explaining the data—and a constraint-based approach—where conditional independence tests are used to determine whether a direct causal relationship should be postulated between two variables. Many variants of these techniques have been applied to gene regulatory network learning, including search-based approaches [6,7,8], information-theoretic approaches [9], parameter-tying based approaches [10], and conventional dynamic Bayesian network learning approaches [11,12].

Although these previous techniques have achieved some promising results, the fundamental limitation of the amount of data available relative to the large number of parameters estimated (e.g. distinct parameters used to predict the expression level of each gene given other genes) severely constrains their effectiveness. This difficulty is inherent to the task: orders of magnitude more expression data would be required for naive estimation approaches devoid of background knowledge and biologically relevant assumptions to succeed on this problem.

One common shortcoming in the current literature, whether using linear modeling or using Bayesian network structure learning, is that nearly all proposed approaches attempt to determine the regulation structure for each target gene independently. Yet it is well known that genes that share the same expression pattern are likely to be involved in the same regulatory process, and therefore share the same (or at least a similar) set of regulators [13]. The main question we investigate is how to exploit biologically significant knowledge about co-regulation to improve the inference of the underlying gene regulatory network from expression data. Although a few previous investigators, such as van Someren et al. (2000), have proposed to group genes with similar expression profiles in a single prototypical "gene", and then model the relations between prototypical genes instead of modeling the genes individually, this is a somewhat oversimplified approach that ultimately ignores the individual differences between genes in the same group, and puts a particular high requirement on the clustering step.

In this paper, we propose a novel approach for predicting the regulators for a given group of genes with similar mRNA expression patterns, by minimizing a globally shared regularized prediction risk that encourages similar genes to share regulators. The models we learn, however, are otherwise standard linear models. The novelty of the approach is to first cluster the genes based on their time series expression profiles, and then minimize a loss determined on a set of global indicator variables associated with the common set of possible regulatory variables. We evaluate the performance of our approach on both synthetic data and the cell cycle time-series gene expression data of [14]. Our synthetic results show that our approach is able to learn the correct structure far more effectively than the typical approach that does not take into account co-regulation knowledge. Our results on the Cho et al. (1998) cell cycle data suggests our approach can identify the important transcription factors in the cell cycle genes more accurately by exploiting the co-regulation knowledge.

2 Method

The core of our method is based on using linear regression models to infer the expression level of each target gene from the expression levels of a set of potential regulator genes. However, even though linear prediction provides a simple and elegant foundation for modeling time series expression data, it cannot be applied naively. At least three significant issues need to be addressed before reasonable results can be achieved. First, time lags exist in the regulatory pathways controlling gene expression. These time lags vary between pathways and remain generally unknown *a priori* [12]. Second, the number of parameters required by a simple linear model (one parameter for each target-regulator combination) is far too many to be estimated reliably from available time series gene expression data. Some sort of effective feature selection mechanism must be employed [5]. Third, genes that serve related or synchronized functions tend to share common regulatory mechanisms. That is, related genes tend to share common regulators, and this knowledge must be exploited somehow to improve the quality of the regulation networks that are inferred. Failure to take into account any of these issues causes the linear prediction (or any other) approach to perform poorly.

We take into account all three of the above issues and modify the linear prediction approach to infer gene regulatory networks from time series expression data. The first two issues have been handled in varying ways in existing research—although we propose particularly simple and elegant ways to handle them in this paper. The third issue comprises the main observation we make, and motivates our use of a novel form of global risk minimization that is able to share regulatory information between similar genes while simultaneously allowing individual differences.

2.1 Linear Modeling

First, to establish the basic linear prediction approach consider an $n \times t$ matrix Y of time series gene expression data, where each column corresponds to the expression levels of a single gene measured over a series of n time points; hence, Y stores the expression profiles for t genes. For each gene, we would like to identify which other genes measured in Y are likely to be regulators. The fundamental hypothesis is that the expression levels of a regulator gene should be predictive of the expression levels for a regulated target gene, possibly subject to time lag and the presence of co-regulators or absence of inhibitors.[1]

A straightforward linear prediction approach proceeds as follows. Assume for a target expression profile $\mathbf{y}_j$ given by an $n \times 1$ column vector from Y, we have a set of candidate regulator profiles stored in an $n \times k$ matrix X_j consisting of k distinct columns selected from Y. (We will discuss below how such a set of candidate profiles might be inferred for a given target $\mathbf{y}_j$.) The quality of this set of candidate regulators can be assessed by how well their expression levels predict the expression levels of the target, which can be determined by solving for the combination weights of the regulator profiles that best reconstruct the target profile

$$\min_{\mathbf{w}_j} \|X_j\mathbf{w}_j - \mathbf{y}_j\|_2^2. \tag{1}$$

Here the $k \times 1$ vector of combination weights $\mathbf{w}_j$ describes how the expression levels of the regulator genes in X_j interact to best explain the target expression levels $\mathbf{y}_j$, and the quality of the fit can be assessed by the residual error in (1).

2.2   Coping with Time Lags via Time Shifting

Unfortunately, the naive linear modeling approach (1) suffers from the three major drawbacks mentioned above. The first problem is that it does not account for any time lag between the expression of a regulating gene and the expression of its downstream target. In fact, the naive approach (1) implicitly assumes that regulation occurs instantaneously, and therefore performs quite poorly at identifying any regulatory relationship that exhibits delayed effects. To cope with this shortcoming, we modify the approach to first take into account any potential time lag between the expression of a regulator and its downstream target. In particular, for each candidate regulator measured in X_j, given by an $n \times 1$ vector $\mathbf{x}_{ij}$, we first compute an optimal shift back in time that best aligns $\mathbf{x}_{ij}$ individually with the target $\mathbf{y}_j$

$$s_{ij}^* = \arg\min_{s \in \{0,1,2,3\}} \|\mathbf{x}_{ij}(1, ..., n-s) - \mathbf{y}_j(s+1, ..., n)\|_2^2. \tag{2}$$

(Note that the shifts only allow time lags forward in time from the expression of the regulator to the expression of the target.) Repeating this for each candidate regulator profile in X_j, yields a series of optimal time lags. We can then reformulate the expression matrix X_j for the candidate regulators by applying the

---

[1] To mitigate the effect of measurement errors and outliers in the expression data, we generally assume the columns of Y have been rescaled to values between 0 and 1, and thus we are only searching for explanations of *relative* increases or decreases in expression level.

optimal shift to each column, and truncating the columns to a common length based on the maximum shift, obtaining an $(n - s_{max}) \times k$ time-lag aligned matrix Φ_j. The target expression profile $\mathbf{y}_j$ is then also truncated to a corresponding $(n - s_{max}) \times 1$ vector $\tilde{\mathbf{y}}_j$, where $\tilde{\mathbf{y}}_j = \mathbf{y}_j(s_{max}, ..., n)$. The quality of the candidate regulators can then be assessed by the more appropriate aligned reconstruction

$$\min_{\mathbf{w}_j} \ \|\Phi_j \mathbf{w}_j - \tilde{\mathbf{y}}_j\|_2^2. \tag{3}$$

2.3 Feature Selection via L1 Regularized Risk Minimization

Although the modified linear approach (3) appropriately handles time lags between regulator and target expression patterns, it still suffers from a major drawback: the set of candidate regulators for a given gene is usually very large (e.g. the complete set of remaining genes), while the number of time points sampled in a time series experiment is usually quite small (on the order of 20 to 30). Therefore a large set of combination weights $\mathbf{w}_j$ need to be inferred from a limited amount of data. Moreover, only a tiny fraction of the candidate regulators are expected to be true regulators for any given gene, meaning that, ideally, most of the weights should be set to 0 to indicate non-regulation. The bottom line is that some sort of effective form of *feature selection* is required for this problem. From a large set of candidate regulator expression profiles, most need to be discarded, and a small number retained to provide a good explanation of the target expression profile.

It is well known in the machine learning literature [15] that using the L1 norm (rather than the more conventional L2 norm) for regularization is very effective for feature selection. In this approach, one adds a penalty to the risk (the reconstruction objective) which encourages small values for $\mathbf{w}_j$:

$$\min_{\mathbf{w}_j} \ \|\Phi_j \mathbf{w}_j - \tilde{\mathbf{y}}_j\|_2^2 + \alpha \|\mathbf{w}_j\|_1, \tag{4}$$

where α is a parameter that trades off the influence of the risk with the regularizer. Crucially, this regularizer encourages many of the weights to become exactly zero in the solution. To see why, note that the regularization term is non-differentiable at zero, but any movement of a weight from zero immediately creates a derivative of magnitude α encouraging movement back to zero. Thus, if the magnitude of the derivative of the risk is not greater than α, then the weight will remain at zero. These intuitions lead to an efficient optimization procedure known as grafting [16].

2.4 Regulation Sharing via Globally Regularized Risk Minimization

Simply solving the minimization problem in (4) provides no advantage over the approaches proposed in the literature however, since it does not address the problem of facing a shortage of data while trying to make inferences about a large number of genes. To mitigate this problem we propose to share regulatory information across sets of target genes. Given the hypothesis that genes with similar expression patterns are usually co-regulated and involved in the same functional process, we propose to first cluster the target genes based on their expression patterns. (This clustering can be performed in many different ways. In our implementation below we simply used a straightforward K-means method.) Then, for each cluster, our goal is to identify a set of regulators that is shared among the entire set of genes in the cluster, while still allowing for differences among the regulation of individual genes. Achieving this type of information sharing in the context of regularized linear modeling (4) however, requires some novel technical developments.

In [17] we recently developed a novel convex Bayesian network structure learning approach based on introducing a set of auxiliary indicator variables to control global feature selection. Adapting this idea to the current context, we propose to use a global regularization scheme on auxiliary selection variables to help identify the common candidate regulators among a group of target genes with similar expression profiles. Given that there is much more data available for sets of similar genes, as opposed to individual genes, we hope that the common regulators can be more accurately identified.

Specifically, given a set of target genes $Y = \{\mathbf{y}_1, ..., \mathbf{y}_m\}$, we would like to identify a common set of regulators from the set of candidates $X = \{\mathbf{x}_1, ..., \mathbf{x}_l\}$. Define a set of indicator variables $\boldsymbol{\eta} = \{\eta_1, ..., \eta_l\}^\top$, corresponding to the candidate set $X = \{\mathbf{x}_1, ..., \mathbf{x}_l\}$, such that each $\eta_i \in \{0,1\}$ indicates whether a regulator X_i is selected as an active regulator. Let $N = \mathrm{diag}(\boldsymbol{\eta})$. Then, we can form a globally regularized version of the minimization problem (4) by introducing the selection variables $\boldsymbol{\eta}$ and adding a new global regularization term on these variables:

$$\min_{\boldsymbol{\eta} \in \{0,1\}^n} \min_{\mathbf{w}} \sum_j \left(\|\Phi N \mathbf{w}_j - \tilde{\mathbf{y}}_j\|_2^2 + \alpha \|\mathbf{w}_j\|_1 \right) + \lambda \mathbf{u}^\top \boldsymbol{\eta}, \tag{5}$$

where $\mathbf{u}$ is a positive weight vector that allows one to incorporate prior knowledge about the importance of each global feature. Although we simply set this vector to 1 in our later experiments, it will be very useful whenever prior knowledge is available. Note that the global regularization term $\lambda \mathbf{u}^\top \boldsymbol{\eta}$ is in fact an L0 norm regularizer, which will automatically force a sparse solution that selects only a small set of global features for the set of target genes in a cluster. Nevertheless, the local L1 norm regularizer, $\alpha \|\mathbf{w}_j\|_1$, will still make individual choices of regulators for each specific target gene; choosing these regulators from the globally selected features identified by $\boldsymbol{\eta}$. Therefore, if the target genes in a cluster share some common regulators, the global feature selection process will be very helpful to pick them out, while the ability to individually model the regulation of each gene has not been diminished.

2.5   Optimization Procedure

Equation (5) encodes a *min-min* integer optimization problem. Unfortunately, integer optimization problems of this form are generally NP-hard. To attempt to solve the problem efficiently, we first relax it into an optimization over continuous variables, by relaxing each $\eta_i \in \{0, 1\}$ to be continuous $\eta_i \in [0, 1]$. This leads to solve the following relaxed *min-min* optimization:

$$\min_{\boldsymbol{\eta}} \min_{\mathbf{w}} \sum_{j} \left(\|\Phi N \mathbf{w}_j - \tilde{\mathbf{y}}_j\|_2^2 + \alpha \|\mathbf{w}_j\|_1 \right) + \lambda \mathbf{u}^\top \boldsymbol{\eta}$$

$$s.t. \quad 0 \leq \boldsymbol{\eta} \leq 1. \tag{6}$$

In fact, this formulation has relaxed the original L0 norm regularizer over $\boldsymbol{\eta}$ into a L1 norm regularizer. In this way we maintain feature selection ability, while gaining computational efficiency.

In our implementation below, we conduct the optimization in two alternating steps: $\min_{\mathbf{w}}$ and $\min_{\boldsymbol{\eta}}$. Each $\min_{\mathbf{w}}$ step is simply a minimization of least squares regression error with L1 norm regularization, which can be implemented as a quadratic program [18], or by using a fast grafting algorithm [16]. For the $\min_{\boldsymbol{\eta}}$ step, we use a quasi-Newton BFGS method to perform the optimization [19].

3   Experiments and Results

We conducted experiments on both synthetic and real cell cycle data to evaluate our approach. In particular, we compared our global regularization approach to the standard independent local predication approach, and a prototype based linear regression method adapted from [20]. Synthetic experiments are useful to gauge the potential effectiveness of the approach under controlled conditions where the ground truth is available. Once the intuitive behavior of the technique is understood, we then apply the method to inferring the structure of the regulatory network of the yeast cell cycle.

In our experiments, we assume all transcription regulations work through activators, instead of inhibitors; that is, we assume the $\mathbf{w}$ parameters are non-negative in the linear regressions. Also, to keep the $\mathbf{w}$ parameters from becoming too small and causing a threshold selection problem, we included the additional constraint $\|\mathbf{w}_j\|_1 \geq 1$ in the three linear regression algorithms.

3.1   Experiments on Synthetic Data

For the synthetic experiments, we set up a small system to simulate a cell cycle process controlled by a small number of critical transcription factors (TFs). We defined 10 TFs that regulated the expression levels of 212 genes in 4 phases of a synthetic cell cycle. These 10 TFs were divided into 4 regulatory groups, with 3, 2, 3, and 2 TFs in each group respectively. Each group of TFs was associated with a specific phase of the cell cycle, and regulated the expression of 53 genes, as well as the TFs in the next phase of the cycle. In our setting, we assumed that one gene (including the TFs themselves) can be regulated by either one TF or a combination of two TFs. We generated the expression data by first simulating ideal expression levels for the TFs in a selected phase for two complete cell cycles, totaling 16 time steps. Then we generated the expression profiles of the genes (or TFs) in the next phase by a 2 time step delayed response from the combination ("and") of m $(m \leq 2)$ randomly selected TFs in their previous phase, plus Gaussian noise. Repeating this procedure for all the phases in the cycle in turn, we generated synthetic time-series profiles for the entire set of TFs and genes.

Both our global regularization approach and the prototype based method require the genes to first be clustered based on their expression profiles. Although the number of clusters used has a minor effect on the performance of both algorithms, the impact is not significant provided that the cluster number is not extreme (neither extremely big nor extremely small). For our synthetic experiments, we simply choose to use 10 as the number of clusters.

Column 5 in Figure 1 shows the expression profiles for the genes and TFs after their profiles have been clustered into 10 groups. We then learn the regulators for the genes in each group, using our globally regularized linear regression to encourage genes in the same group to share parents. We compared the results of the global approach to both the standard "local" approach of learning the parent regulators for each gene separately, and the prototype based approach of forcing all the genes in one group to have the exactly same set of parents. The comparison algorithms serve as controls at the two opposite extremes. We used the same L1 regularized method for parent selection in all of the algorithms. After obtaining the $\mathbf{w}$ parameters from each algorithm, all the parents indicated by $\mathbf{w} > 10^{-5}$ are determined as predicted regulators for the corresponding genes. For a fair comparison, the regularization parameters (α and λ) were chosen to yield the highest F-measure values in each case.

Columns 1–3 in Figure 1 show the regulator prediction results for the three algorithms respectively; comparing them with the true regulation information in Column 4. The x-axis for each column indicates the candidate TFs from which a subset is selected as the set of regulators for each gene. The y-axis for each column indexes the individual target genes. Each row plots the predicted regulators for each gene based on the corresponding $\mathbf{w}$ parameters for that gene, where white indicates a large value (indicating a regulator), while dark indicates a value close to 0 (indicating no regulation).

The table in Figure 1 compares the performance of the three algorithms. The *precision* score measures true positive predictions (tp) divided by true positives plus false positive predictions (fp). That is, $precision = tp/(tp+fp)$. Similarly, *recall* score is measured in terms of the number of false negative predictions (fn), and is given by $recall = tp/(tp + fn)$. *F-measure* is a standard combination of both precision (p) and recall (r), given by $F\text{-}measure = 2\,p\,r/(p+r)$. The *accuracy* score measures the proportion of the correct predictions. That is, $accuracy = (tp + tn)/(tp + tn + fp + fn)$. Here we can see that the global regularization approach greatly outperforms both the local regularization and prototype based methods with respect to both accuracy and F-measure. The local predication method is not able to effectively identify the true regulators due to the noise in the data and the limited number of time points. The prototype base method

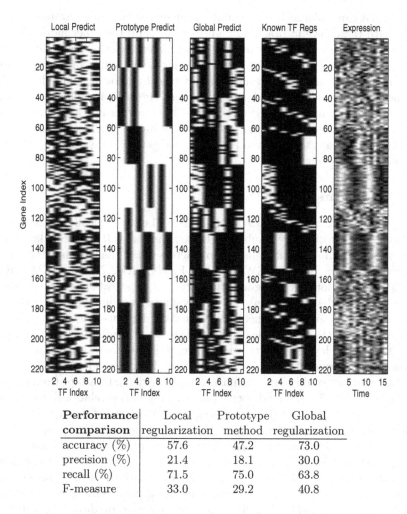

Fig. 1. Results on synthetic data. Rows denote target genes in the synthetic experiment. Columns denote candidate regulators (transcription factors). A white cell denotes a large weight ($w_{ij} > 10^{-5}$) connecting a TF j to a target gene i in the estimated linear model, indicating that j is inferred to regulate i. A black cell denotes a small weight ($w_{ij} \leq 10^{-5}$), indicating that j is not inferred to regulate i. Column 1: local prediction output. Column 2: prototype prediction output. Column 3: global prediction output. Column 4: ground truth regulatory relationships. Column 5: expression level data used as input.

also has difficulty identifying correct regulatory relationships, and tends to choose too many parents for each gene. The reason for this is clear however. Since the prototype method is forced to choose a single set of regulators for controlling a large set of genes, it naturally chooses the union of the prospective regulators for each gene, leading to subsequently low precision and accuracy. Thus, the prototype approach depends heavily on having a more refined and accurate set of clusters from which it can make accurate regulatory inferences, but an accurate clustering is very hard to achieve in practice. Figure 1 shows, on the other hand, that the global regularization approach can effectively remove irrelevant candidate TFs by sharing co-regulation information within a group, while simultaneously reducing the number of spurious regulators being inferred by allowing individual differences between genes in a given cluster. The overall result is a much more accurate (albeit far from perfect) recovery of the underlying regulatory structure.

The main question that remains is whether the higher quality inference on this synthetic model leads to improved results on real gene expression data, which we consider next.

3.2   Experiments on Real Data

Gene expression microarray data for the yeast cell cycle typically contains more than 6000 genes, while only a subset of these genes are cell cycle regulated. It is known there are 9 important transcription factors (TFs) that regulate the cell cycle process [21], namely: SWI4, SWI6, MPB1, FKH1, FKH2, NDD1, MCM1, ACE2 and SWI5. Since a lot of gene regulatory relationships have already been identified for yeast, this model is commonly used to evaluate learning approaches that attempt to infer gene regulatory networks from data. Here we use Cho et al.'s data [14], and focus on the task of identifying the subset of regulators from the 9 candidate TFs, for each yeast gene that is cell cycle regulated. To clearly evaluate our approach, we chose a subset of 267 cell cycle regulated genes from the Cho et al. data [14], while we could obtain confirmed regulatory relationships from the previous literature [21,22], or could obtain potential regulation relationships from existing binding data [21] for 127 genes among them. We rescaled the expression data to values between 0 and 1, and then clustered the genes into 15 clusters using K-means. (In the images shown in Figure 2, the genes are grouped vertically into the clusters. The number of clusters is chosen by using visual judgment to achieve a smooth clustering effect.) Finally, we tested our algorithms on each cluster. As in the synthetic experiments, after obtaining the $\mathbf{w}$ parameters from each algorithm, all the parents indicated by $\mathbf{w} > 10^{-5}$ are determined as predicted regulators for the corresponding genes. For a fair comparison, the regularization parameters (α and λ) were chosen to yield the highest F-measure values in each case.

Since the regulatory mechanisms are still not known for a portion of the 267 genes, we therefore can only evaluate the results over the 127 genes for which regulatory relationships are presumed known. Figure 2 shows the prediction results on 127 genes for all the three algorithms: locally regularized prediction, prototype based prediction, and globally regularized prediction. The images compare the performance of the three methods on inferring regulators from among the 9 candidate TFs, and shows how they related to the known TF-based regulatory relationships. These results show that the globally regularized approach can significantly improve the quality of both the standard locally regularized approach and the prototype based approach adapted from [20]. As in the synthetic case, the globally regularized approach has the ability to share regulatory information between genes within a cluster, leading to better noise robustness than the

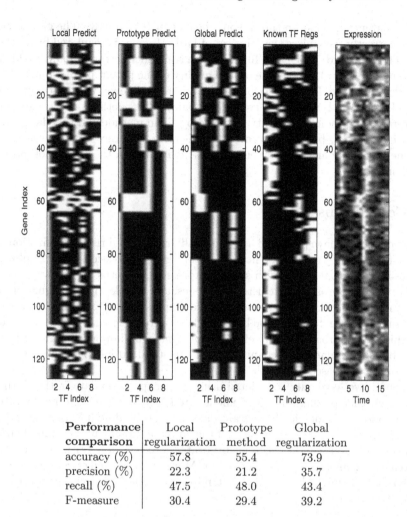

Fig. 2. Results on the subset of the real gene expression data from [14], restricted to genes where TF-based regulation information is known or can be inferred from other sources [21,22]. Rows denote target genes in the synthetic experiment. Columns denote candidate regulators (transcription factors). A white cell denotes a large weight ($w_{ij} > 10^{-5}$) connecting a TF j to a target gene i in the estimated linear model, indicating that j is inferred to regulate i. A black cell denotes a small weight ($w_{ij} \leq 10^{-5}$), indicating that j is not inferred to regulate i. Column 1: local prediction output. Column 2: prototype prediction output. Column 3: global prediction output. Column 4: ground truth regulatory relationships. Column 5: expression level data used as input.

local approach. Here too, the global technique also overcomes the problem of being overly dependent on clustering quality, like the prototype approach, by allowing regulation differences with a cluster. For example, in Figure 2, in the group of genes indexed between 42-58, one can see that a large set of the errors produced by the standard independent approach (Column 1) have been corrected by sharing parent information throughout the cluster (Column 3). The global regularizer correctly recognizes that this set of late-G1 genes is regulated by a subset of SWI4/SWI6 and MBP1/SWI6. Although some local errors remain in this region (and elsewhere), clearly the overall quality of the parent prediction has been improved substantially in the global method. For these genes, the prototype based method (Column 2) recognizes two additional parents, perhaps due to noise.

Overall, the prediction quality achieved by these methods on this data is still somewhat limited, but has improved significantly over the past few years, and in some sense is remarkable given the noise exhibited in the expression profiles (Column 5).

4 Conclusions

In this paper, we have proposed a new globally regularized risk minimization objective for learning regulatory networks from gene expression data. Exploiting the assumption that genes with similar expression patterns are likely to be co-regulated, our approach first clusters the genes, and then learns the regulatory relationships by encouraging genes with similar expression patterns to share regulators. Our experimental results on both synthetic data and real cell cycle data show that this new approach is more effective at identifying important (transcription factor based) regulatory mechanisms than the standard independent approach, and a prototype based approach.

Thus far, we have only considered using gene expression data in the learning process. Further prediction improvements are likely to come from incorporating further sources of biologically relevant data, such as binding information [21], or other forms of prior knowledge beyond the co-regulation assumption made here. These informations can be nicely incorporated into our global risk minimization approach by using the **u** parameter vector. Moreover, as an effective feature selection strategy, it might be useful to extend this approach resolving other feature selection bioinformatics problems.

Acknowledgments

Research supported by NSERC, MITACS, CFI, the Alberta Ingenuity Centre for Machine Learning, and the Canada Research Chair program. We thank the anonymous referees for their helpful comments.

References

1. D'Haeseleer, P., Wen, X., Fuhrman, S., Somogyi, R.: Linear modeling of mrna expression levels during cns development and injury. Pac. Symp. Biocomput., 41–52 (1999)
2. Chen, K.C., Wang, T.Y., Tseng, H.H., Huang, C.Y.F., Kao, C.Y.: A stochastic differential equation model for quantifying transcriptional regulatory network in saccharomyces cerevisiae. Bioinformatics 21, 2883–2890 (2005)

3. De Jong, H., Gouze, J.L., Hernandez, C., Page, M., Sari, T., Geiselmann, J.: Qualitative simulation of genetic regulatory networks using piecewise-linear models. Bull. Math. Biol. 66, 301–340 (2004)
4. De Hoon, M., Imoto, S., Kobayashi, K., Ogasawara, N., Miyano, S.: Inferring gene regulatory networks from time-ordered gene expression data of bacillussubtilis using differential equations. Pac. Symp. Biocomput., 17–28 (2003)
5. Li, F., Yang, Y.: Recovering genetic regulatory networks from micro-array data and location analysis data. Genome Informatics 15, 131–140 (2004)
6. Hartemink, A., Gifford, D., Jaakkola, T., Young, R.: Using graphical models and genomic expression data to statistically validate models of genetic regulatory networks. Pac. Symp. Biocomput., 422–433 (2001)
7. Yu, J., Smith, V., Wang, P., Hartemink, A., Jarvis, E.: Advances to Bayesian network inference for generating casual networks from observational biological data. Bioinformatics 20, 3594–3603 (2004)
8. Wang, S.: Reconstructing genetic networks from time ordered gene expression data using Bayesian method with global search algorithm. J. Bioinform. Comput. Biol. 2, 441–458 (2004)
9. Chen, X., Anantha, G., Wang, X.: An effective structure learning method for constructing gene networks. Bioinformatics 22, 1367–1374 (2006)
10. Segal, E., Pe'er, D., Regev, A., Koller, D., Friedman, N.: Learning module networks. J. Mach. Learn. Res. 6, 557–588 (2005)
11. Bernard, A., Hartemink, A.: Informative structure priors: joint learning of dynamic regulatory networks from multiple types of data. Pac. Symp. Biocomput., 459–470 (2005)
12. Zou, M., Conzen, S.: A new dynamic Bayesian network (DBN) approach for identifying gene regulatory networks from time course microarray data. Bioinformatics 21, 71–79 (2005)
13. D'Haeseleer, P., Wen, X., Fuhrman, S., Somogyi, R.: Genetic network inference: from co-expression clustering to reverse engineering. Bioinformatics 16, 707–726 (2000)
14. Cho, R.J., Campbell, M.J., Winzeler, E.A., Steinmetz, L., Conway, A., Wodicka, L., Wolfsberg, T.G., Gabrielian, A.E., Landsman, D., Lockhart, D.J., Davis, R.W.: A genome-wide transcriptional analysis of the mitotic cell cycle. Mol. Cell. 2, 65–73 (1998)
15. Ng, A.: Feature selection, L1 vs L2 regularization, and rotational invariance. In: International Conf. on Mach. Learn (ICML) (2004)
16. Simon, P., Kevin, L., James, T.: Grafting: Fast, incremental feature selection by gradient descent in function space. J. Mach. Learn. Res. 3, 1333–1356 (2003)
17. Guo, Y., Schuurmans, D.: Convex structure learning for Bayesian networks: Polynomial feature selection and approximate ordering. In: Conf. on Uncertainty in Artif. Intell (UAI) (2006)
18. Boyd, S., Vandenberghe, L.: Convex Optimization. Cambridge Univ. Press (2004)
19. Bertsekas, D.: Nonlinear Optimization. Athena Scientific (1995)
20. van Someren, E., Wessels, L., Reinders, M.: Linear modeling of genetic networks from experimental data. Intelligent Systems for Molecular Biology (ISMB 2000), 355–366 (2000)
21. Simon, I., Barnett, J., Hannett, N., Harbison, C., Rinaldi, N., Volkert, T., Wyrick, J.J., Zeitlinger, J., Gifford, D., Jaakkola, T., Young, R.: Serial regulation of transcriptional regulators in the yeast cell cycle. Cell 106, 697–708 (2001)
22. Iyer, V.R., Horak, C.E., Scafe, C.S., Botstein, D., Snyder, M., Brown, P.O.: Genomic binding sites of the yeast cell-cycle transcription factors sbf and mbf. Nature 409, 533–538 (2001)

Evolution of Tandemly Arrayed Genes in Multiple Species

Mathieu Lajoie, Denis Bertrand, and Nadia El-Mabrouk

DIRO - Université de Montréal - H3C 3J7 - Canada
{lajoimat,bertrden,mabrouk}@iro.umontreal.ca

Abstract. Tandemly arrayed genes (TAG) constitute a large fraction of most genomes and play important biological roles. They evolve through unequal recombination, which places duplicated genes next to the original ones (tandem duplications). Many algorithms have been proposed to infer a tandem duplication history for a TAG cluster in a single species. However, the presence of different transcriptional orientations in most TAG clusters highlight the fact that processes such as inversions also contribute to their evolution. This makes those algorithms unsuitable in many cases. To circumvent this limitation, we proposed in a previous work an extended evolutionary model which includes inversions and presented a branch-and-bound algorithm allowing to infer a most parsimonious scenario of evolution for a given TAG cluster. Here, we generalize this model to multiple species and present a general framework to infer ancestral gene orders that minimize the number of inversions in the whole evolutionary history. An application on a pair of human-rat TAG clusters is presented.

1 Introduction

A multigene family is a set of genes that have evolved by duplication from a common ancestral gene, and share a similar sequence and usually a similar function. Members of a gene family in a given genome may appear in clusters, or scattered in a single or many chromosomes. In this paper, we focus on families of tandemly arrayed genes (TAG): copies that are adjacent on the chromosome. TAGs have been shown to represent a large proportion of genes in a genome. In particular, they represent about 14-17% of all genes in human, mouse and rat [26]. Clusters of TAGs may vary in length from two to hundreds genes, though small clusters are largely predominant (an average of 3 to 4 genes in mouse, rat and human) [26]. They are involved in many different functions of binding or receptor activities. In particular, the olfactory receptor genes constitute the largest multigene family in the vertebrate genome, with several hundred genes per species [1]. Other families of TAGs include the HOX genes [31], the immunoglobulin and T-cell receptor genes [2], the MHC genes [17] and the ZNF genes encoding for transcription factors [25].

TAGs are widely viewed as being generated solely by tandem duplications resulting from unequal recombination [15] or slipped strand mispairing [3]. Such

G. Tesler and D. Durand (Eds.): RECOMB-CG 2007, LNBI 4751, pp. 96–109, 2007.

mechanisms have the effect of generating sequences of repetitive units with the same transcriptional orientation. However, it is not infrequent to observe TAGs with different orientations. In particular, Shoja and Zhang [26] have observed that more than 25% of all neighboring pairs of TAGs in human, mouse and rat have non-parallel orientations. This underlines the fact that other rearrangement mechanisms such as inversions should be considered in an evolutionary model of TAGs.

Based on the unequal recombination model of evolution, a large number of studies have considered the problem of reconstructing a tandem duplication history of a TAGs family [5,11,12,27]. These are essentially phylogenetic inference methods using the additional constraint that the resulting tree should induce a duplication history according to the given gene order. When a gene tree is already available for a gene family, a linear-time algorithm can be used to check whether it is a duplication tree [32]. However, it is often impossible to reconstruct a duplication history [16], due to other evolutionary events such as gene losses or genomic rearrangements [10]. In [6,21] we have considered an evolutionary model accounting for both tandem duplications and inversions. Given a gene tree for a family of TAGs, we developed an algorithm allowing to find the minimum number of inversions in any possible evolutionary scenario for this family.

All the above methods are restricted to the analysis of TAGs located on a single chromosome (and thus in a single species). However, the increasing availability of complete genomic sequences and of many different TAGs databases [1,29] makes it possible to study the evolution of gene families with members belonging to different species. Such a global evolutionary study may help deciphering the common origins of TAGs, highlighting the inter-species differences and identifying the genetic basis of species-specific features. Various phylogenetic studies have been conducted by biological groups on different TAGs families such as the Zinc-Finger transcription factors in human and mouse [25], and the olfactory receptor genes in various mammalian species [1]. However no rigorous approach has been developed so far to explain the non agreement between a given gene tree of a TAGs family and a duplication and speciation history.

In this paper, we consider an evolutionary model of TAGs accounting for duplication, speciation, gene loss and inversion events. This is a generalization of [21] to TAGs located on different genomes. More precisely, given a gene tree for a family of TAGs and their signed order on the genomes (chromosomes or clusters), we aim to find an evolutionary scenario involving the minimum number of inversions, and the corresponding gene orders of the ancestral genomes. The Fitch model allows for the simultaneous duplication of several gene copies, but there are now evidence that simple duplications are predominant over multiple duplications [5,31]. As a first attempt, we only consider simple duplications.

This paper is organized as follows. After describing the evolutionary models in Section 2, we present the general problem in Section 3. It is related to the more classical one of inferring the gene order of the ancestral genomes in a species tree minimizing a given genomic distance [23,24]. In Section 3.1, we present an algorithm to infer the most parsimonious scenario of inversion on a single branch of the species tree. In Section 3.2, we present a simple iterative method used to infer the ancestral gene orders minimizing the total number of inversions in a species tree. It is based on the median problem, for which we propose a branch-and-bound algorithm in Section 3.3. Finally, in Section 4, we test the algorithm's time-efficiency on simulated data, and present an application on a pair of human-rat TAGs clusters.

2   The Evolutionary Model

The classical model of evolution considered for TAGs is based on tandem duplications resulting from unequal recombination during meiosis, which is assumed to be the sole evolutionary mechanism (except point mutations) acting on sequences. Formally, from a single ancestral gene at a given position in the chromosome, the locus grows through a series of consecutive duplications placing the created copy next to the original one. Such *tandem duplications* may be *simple* (duplication of a single gene) or multiple (simultaneous duplication of neighboring genes). In this paper, we only consider simple duplications. From now on, a *duplication* will refer to a simple tandem duplication.

This model of evolution applies only to family of TAGs all located on the same chromosome and having the same transcriptional orientation. In particular, it is unsuitable to describe the evolution of a TAGs family containing members on both DNA strands. To circumvent this limitation, we have proposed, in [21], an extended model of duplications including inversions. In this paper, we further extend the model to account for several genomes.

Consider a family of TAGs located on m different genomes. In addition to the TAGs orders on each genome, all we can infer from the gene sequences is a gene tree representing the global evolution of the gene family. Formally, an *ordered gene tree* is a set $(T, \mathcal{O})$, where T is a gene tree of the TAGs family and $\mathcal{O} = (O_1, O_2, \cdots O_m)$ where O_i is the signed order of the family members in genome i, for $1 \leq i \leq m$. Thereafter, the transcriptional orientations of the genes in an ordered gene tree $(T, \mathcal{O})$ are specified by signs $(+/-)$ in each O_i. We denote by $d_{inv}(O_i, O_j)$ the inversion distance between the two signed permutations O_i and O_j. Such a distance can be computed using the original Hannenhalli and Pevzner algorithm[18], or any of the existing optimizations [4,20,28].

An ordered gene tree $(T, \mathcal{O})$ can always be explained by a history $\mathcal{H}$ involving duplication, gene loss, inversion and speciation events (DLIS history), as stated in Lemma 1 below. We say that $\mathcal{H}$ is a DLIS history of $(T, \mathcal{O})$. Hereafter, we begin by formally defining a DLIS history (see Figure 1 for an illustration).

Definition 1. *Let $\mathcal{H} = ((T^1, \mathcal{O}^1), \cdots (T^k, \mathcal{O}^k), \cdots (T^{n-1}, \mathcal{O}^{n-1}), (T^n, \mathcal{O}^n))$ be a sequence of n ordered gene trees. For each k, $1 \leq k \leq n$, we denote by m_k the number of genomes represented in T^k, and by O_i^k the gene order in genome i, for $1 \leq i \leq m_k$.*

We say that $\mathcal{H}$ is a DLIS history if and only if:

1. *$T^1 = v$ is the single leaf gene tree and $\mathcal{O}^1 = (O_1^1) = (\pm v)$ is one of the two trivial orders;*

2. *For $1 < k < n$, one of the four following situations hold:*

    a. Duplication event: *There is an i, $1 \le i \le m_k$, such that T^{k+1} is obtained from T^k by adding two children u and w to a leaf v belonging to genome i. Moreover:*

        – $m_{k+1} = m_k$;

        – $\mathcal{O}^{k+1}$ *is obtained from $\mathcal{O}^k$ by replacing $\pm v$ by $(\pm u, \pm w)$ in O_i^k.*

    b. Gene loss event: *There is an i, $1 \le i \le m_k$, such that T^{k+1} is obtained from T^k by removing a leaf v belonging to genome i. Moreover, if v was the only leaf in O_i^k, $m_{k+1} = m_k - 1$. Otherwise, O_i^{k+1} is obtained from O_i^k by deleting v and $m_{k+1} = m_k$.*

    c. Inversion event: *$T^{k+1} = T^k$ and there is an i, $1 \le i \le m_k$, such that $d_{inv}(O_i^k, O_i^{k+1}) = 1$. Moreover, $O_j^{k+1} = O_j^k$ for $j \ne i$ and $m_{k+1} = m_k$.*

    d. Speciation event: *There is an i, $1 \le i \le m_k$, such that T^{k+1} is obtained from T^k by adding two children u and w to each leaf v belonging to genome i. Moreover:*

        – $m_{k+1} = m_k + 1$;

        – $\mathcal{O}^{k+1}$ *is obtained from $\mathcal{O}^k$ by duplicating the order O_i^k.*

Any DLIS history $\mathcal{H}$ of $(T, \mathcal{O})$ induces a unique species tree S obtained from the speciation events of $\mathcal{H}$. We say that $\mathcal{H}$ is *consistent with* S (see Figure 1).

Let $(T, \mathcal{O})$ be an ordered gene tree for a family of TAGs on m genomes and suppose that the species tree S is known (otherwise we can take advantage of our algorithm presented in [8]). Then a natural problem is to find a

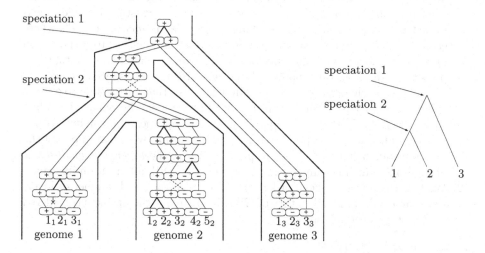

Fig. 1. A DLIS history. Transcriptional orientations are indicated by signs, duplications by bold lines, gene losses by 'X' and inversions by dashed lines. The resulting TAGs orders are denoted as i_k meaning "gene i in genome k". For clarity, we omitted successive identical configurations in each lineage. The induced species tree for the three genomes is given right.

DLIS history of $(T, \mathcal{O})$ that is consistent with S. The existence of such a history is stated in Lemma 1 below. It follows from the existence of a duplication/speciation/loss history in the more general case of non-ordered gene families generated by general duplications, e.g. not necessarily in tandem. More precisely, given a gene tree T for a set of (unsigned) genes located on m genomes, and a species tree S for these genomes, the classical *reconciliation* approach [9,13,22] infers a duplication/speciation/loss history, involving a minimum number of gene losses/duplications, that has led to the gene tree T.

Lemma 1. *Given an ordered gene tree $(T, \mathcal{O})$ on m genomes and a species tree S for the m genomes, there is at least one DLIS history $\mathcal{H}$ of $(T, \mathcal{O})$ consistent with S.*

Proof. Obtain a sequence of duplications (not necessarily in tandem), gene losses and speciations from the reconciliation of T and S. From that sequence, construct a DLIS history $\mathcal{H}' = ((T^1, \mathcal{Q}^1), \cdots, (T^n = T, \mathcal{Q}^n))$ by applying the corresponding rules in Definition 1 (case *a.*, *b.* or *d.*). Then, obtain $\mathcal{H}$ from $\mathcal{H}'$ by appending the inversions required to transform $\mathcal{Q}^n$ in $\mathcal{O}$ (case *c.* in Definition 1)   □

3   An Inference Problem

As the number of possible DLIS histories of $(T, \mathcal{O})$ consistent with S can be huge, we restrict ourselves to finding a most parsimonious one. The fact that T and S are not affected by the inversion events allows us to proceed in two steps:

1. We infer the minimal number of gene losses/duplications and their localization from the reconciliation of S and T, using the existing methods mentioned in the preceding section. This leads to a *reconciled* tree T', where each internal node has two children and is associated either to a speciation or a duplication event. Each speciation node corresponds to a gene that was present in an ancestral genome before the corresponding speciation event.

   This reconciled tree T' may contain new leaves that correspond to extinct genes. Those leaves are irrelevant to our study and can be removed, leading to a new tree T'' that may contain a number of speciation nodes with a single child. They correspond to ancestral genes that were lost in one of the two descendant lineages. An example is shown in the *ordered reconciled tree* of Figure 2, which is simply a reconciled tree T'' with an order on its leaves.

2. We find the ancestral gene orders that minimize the total number of inversions involved in a DLIS history of $(T'', \mathcal{O})$. Formally, the problem considered in this step is the following:

MINIMUM-DLIS PROBLEM
Input: An ordered reconciled tree $(T, \mathcal{O})$.
Output: A gene order for each ancestral genome inducing a DLIS history of minimum inversions.

From the general case of Definition 1, we now introduce the following restricted evolutionary history: a *Duplication/Inversion* history (*DI history*) is

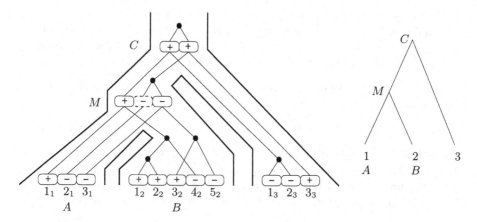

Fig. 2. The ordered reconciled tree induced by the DLIS history of Figure 1, with the corresponding gene order at each ancestral genome preceding each speciation event. The gene tree is "embedded" in the species tree. The dashed gene in genome M has no descendants in lineage B, indicating a gene loss; black dots represent gene duplication.

an evolutionary history of TAGs in one species involving only duplication and inversion events (case $a.$ and $c.$ of Definition 1).

Suppose now we are given an ordered reconciled tree $(T, \mathcal{O})$ with an arbitrary gene order for every internal node of S. There exists a DI history with a minimum number of inversions for each branch of S, and the minimum number of inversion in any DLIS history explaining $(T, \mathcal{O})$ (and the ancestral gene orders) is the sum of the inversions involved in those minimal DI histories. The next section focuses on a single DI history.

3.1   The Generalized Minimum-DI Problem

This problem is a generalization of the Minimum-DI problem we presented in [21], which consists in finding the minimum number of inversions required to explain a single "rooted" ordered gene tree. Here the goal is to find the minimum number of inversions required to explain an *ordered forest of gene trees*, associated to a given branch of the species tree (see Figure 2 and 3a). Formal definitions follow.

Definition 2. *An* ordered forest of gene trees (F, R, O) *is a set of n gene trees $F = \{T_1, T_2, \cdots T_n\}$ rooted at $R = \{r_1, r_2 \cdots r_n\}$ (r_i is the ancestral gene that gave rise to T_i) with an order O on the set of leaves of F. When an ancestral order O_R is imposed on R, we use the notation (F, O_R, O).*

Definition 3. *Let $O_R = (r_1, r_2 \cdots r_n)$ be an ordered sequence of roots, and $O_1 = (o_1, o_2, \cdots o_n)$ be an ordered sequence of genes such that, for each $1 \leq i \leq n$, o_i is a direct descendant of r_i. A partial DI history rooted at O_R is a sequence of ordered forests of gene trees $\mathcal{H} = ((F_1, O_R, O_1), ..., (F_{k-1}, O_R, O_{k-1}), (F_k, O_R, O_k))$ where (F_1 is just a set of single leaf gene trees, and for $0 < i < k$:*

1. Inversion event: If $F_{i+1} = F_i$, then $d_{inv}(O_i, O_{i+1}) = 1$.
2. Duplication event: If $F_{i+1} \neq F_i$, then F_{i+1} is obtained from F_i by adding two children u and w to one of its leaf v, and O_{i+1} is obtained from O_i by replacing v by (u, w), where u and w have the same sign as v.

Moreover, a partial duplication history *is a partial DI history restricted to duplication events.*

A partial duplication history gives rise to a duplication forest, defined as follows.

Definition 4. *A* duplication forest *is an ordered forest of gene trees* (F, O_R, O) *which contains only duplications trees, and such that for every pair of roots* r_i, r_j *in* R, *if* r_i *precedes* r_j *in* O_R, *then all the leaves of* T_i *precedes all the leaves of* T_j *in* O. *Moreover, the leaves of each* T_k *in* F, *must have the same sign as* r_k.

The following theorem is a generalization of the result obtained in [21] for a single ordered gene tree.

Theorem 1. *Let* (F, O_R, O) *be an ordered forest of gene trees and* (F, O_R, O') *be a duplication forest such that* $d_{inv}(O, O') = i$ *is minimum. Then there exists a partial DI history of* (F, O_R, O) *with exactly* i *inversions. Moreover,* i *is the minimum number of inversions involved in any partial DI history of* (F, O_R, O).

Proof. The proof uses arguments similar to those considered in [21], and will be detailed in a full version of this extended abstract   □

Theorem 1 allows us to formulate the problem as follows:

GENERALIZED-MINIMUM-DI PROBLEM
Input: An ordered forest of gene trees (F, O_R, O),
Output: An order O' on the leaves of F such that (F, O_R, O') is a duplication forest and $d_{inv}(O, O')$ is minimal.

For a branch represented by the forest (F, O_R, O), we denote by $DI(O_R, O)$ the minimal $d_{inv}(O, O')$ defined above, and we call it the *minimum DI value*.

A Branch-and-Bound Algorithm. The algorithm is a generalization of the one we presented in [21]. Given an ordered gene tree (T, O), the goal was to find an order O' minimizing the distance $d_{inv}(O, O')$ that is *compatible with* T, i.e. such that (T, O') is a duplication tree. As mentioned in [16], the considered duplication trees are equivalent to binary search trees. Therefore, to enumerate all the orders compatible with T, we associated a binary variable b_i to each internal node i of T as follows: each b_i defines an order relation between the left and right descendant leaves of i, i.e. by setting b_i to 0 (respec. 1), we make all the left descendants smaller than the right ones (respec. all left descendants are larger than the right ones). Then an order O' is compatible with T iff it is defined by an assignment of all the binary variables b_i in T, and all its genes have the same sign (+ or −). If n is the number of leaves of T, this leads to 2^n distinct orders O' compatible with T.

To avoid computing $d_{inv}(O, O')$ for each order O', we considered a branch-and-bound strategy. The idea was to compute a lower bound on $d_{inv}(O, O')$ as we progressively define a partial order O^*, by updating the breakpoint graph of (O, O^*) [18]. The b_i values must be defined in a depth-first manner according to T (see [21] for more details).

Generalization to an ordered forest of gene trees (F, O_R, O) is straightforward. Indeed, let $(T_1, T_2, \cdots T_n)$ be the set of trees of F ordered according to the order O_R of their roots. Then an order O' *compatible with* (F, O_R), i.e. such that (F, O_R, O') is a partial duplication tree, is the concatenation of n orders $(o'_1, o'_2, \cdots o'_n)$ such that o'_i is compatible with T_i. Therefore, similarly to the preceding case, an order O' is compatible with (F, O_R) iff it is defined by an assignment of all the binary variables b_i in F, and for each $1 \leq i \leq n$, all the genes belonging to T_i have the same sign as r_i (see Figure 3). The same branch-and-bound strategy can then be used to explore the space of all possible orders.

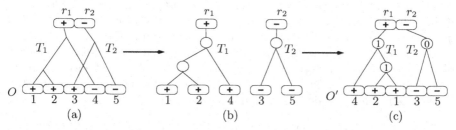

(a)   (b)   (c)

Fig. 3. (a) The ordered forest of gene trees corresponding to the branch (M, B) of the tree in Figure 2 ($F = \{T_1, T_2\}, O_R = (r_1, r_2), O = (1, 2, 3, -4, -5)$). (b) The gene trees in a), with an arbitrary left/right orientation of the children at each internal node. (c) The ordered forest of duplication trees (F, O_R, O') induced by an assignment of the b_i variables in b). The resulting order is $O' = (4, 2, 1, -3, -5)$, and $d_{inv}(O, O') = 3$.

3.2   A General Method Using the Median Problem

The Minimum-DLIS problem is related to the more classical one of inferring the gene orders of the hypothetical ancestral genomes represented by the internal nodes of a species tree. In this case, each species is characterized by a given gene order, and the problem is to find the ancestral gene orders minimizing a given genomic distance. The two distances that have been considered in the literature are the breakpoint and inversion distances [23,24].

Although the case of ordered gene trees is more involved due to the fact that the considered duplication are in tandem, the two problems are related, suggesting a similar global approach summarized below.

1. Begin with an arbitrary order for each internal node of the species tree;
2. Traverse the tree in a depth-first manner. For each subtree consisting of two sister branches (M, A) and (M, B) and a branch (C, M) where C is the

immediate ancestor of M (see Figure 2), ignore the assigned order of M, and reconstruct an order that minimizes the value:

$$DI(O_M, O_A) + DI(O_M, O_B) + DI(O_C, O_M).$$

3. Iterate step 2. a given number of times, or after convergence to a minimizing configuration.

Step 1. can be improved by the use of a heuristic that will be detailed in a full version of this extended abstract. Step 2. can be seen as a generalization of the reversal median problem, which has been proved to be NP-hard [7]. To formally define the median problem, we need to extend the notion of an ordered forest of gene trees by allowing the order to be defined either for the leaves or for the roots of the trees. An ordered forest of gene trees defined by a set of trees F_{XY}, a set of roots X and a set of leaves Y will be denoted as (F_{XY}, X, O_Y) and called a *leaf-ordered forest of gene trees* if an order O_Y is defined on the leaves, by (F_{XY}, O_X, Y) and called a *root-ordered forest of gene trees* if an order O_X is defined on the roots, and by (F_{XY}, O_X, O_Y) and called a *fully-ordered forest of gene trees* if an order is defined for both the leaves and the roots.

The median problem is formulated as follows. Given two leaf-ordered forests of gene trees (F_{MA}, M, O_A) and (F_{MB}, M, O_B) (M is the set of ancestral genes generating both A and B) and a root-ordered forest of gene trees (F_{CM}, O_C, M), the goal is to find an order O_M minimizing the value:

$$DI(O_M, O_A) + DI(O_M, O_B) + DI(O_C, O_M)$$

The following section focuses on the median problem.

3.3   A Branch-and-Bound Algorithm for the Median Problem

To avoid considering each of the $2^n n!$ possible signed orders O_M, where n is the number of genes of M, we consider a branch-and-bound strategy. The idea is to compute a lower bound on $DI(O_M, O_A)$, $DI(O_M, O_B)$ and $DI(O_C, O_M)$ as we progressively extend the prefixes O_M^* of M. This is justified by the following property.

Property 1. Let (F_{XY}^*, O_X^*, O_Y^*) be a fully-ordered forest of gene trees obtained from (F_{XY}, O_X, O_Y) by removing the tree rooted at the last element of O_X, or the leaf corresponding to the last element of O_Y. Then:

$$DI(O_X^*, O_Y^*) \le DI(O_X, O_Y)$$

This bound can be used when we progressively construct the median candidate order O_M. The branch-and-bound strategy is explained below.

1. Consider an initial upper bound for the median problem and the empty orders O_M^*, O_A^*, O_B^* and O_C^*.

Table 1. Average execution time (in seconds) for 1,000 ordered forests of gene trees / Average fraction of the search space explored during the branch-and-bound

	Median size			
	6 genes	8 genes	10 genes	12 genes
4 inversions	$0.18 \ / \ 2 \times 10^{-3}$	$0.24 \ / \ 3 \times 10^{-5}$	$0.30 \ / \ 2 \times 10^{-7}$	$0.54 \ / \ 9 \times 10^{-10}$
6 inversions	$0.48 \ / \ 5 \times 10^{-3}$	$0.84 \ / \ 1 \times 10^{-4}$	$2.10 \ / \ 1 \times 10^{-6}$	$5.70 \ / \ 8 \times 10^{-9}$
8 inversions	$0.90 \ / \ 1 \times 10^{-2}$	$2.76 \ / \ 4 \times 10^{-4}$	$10.92 \ / \ 6 \times 10^{-6}$	$43.38 \ / \ 5 \times 10^{-8}$

2. Construct O_M^* by adding a gene g_M at the end of O_M^*, and construct O_A^* and O_B^* by inserting the genes of O_A and O_B that are descendant of g_M in the right positions. Moreover, if g_M is the descendant of a gene g_C that is not in O_C^*, then construct O_C^* by inserting this gene, otherwise O_C^* is unchanged.
3. Compute $DI(O_M^*, O_A^*)$, $DI(O_M^*, O_B^*)$ and $DI(O_C^*, O_M^*)$, using the branch-and-bound algorithm described in Section 3.1.
4. If $DI = DI(O_M^*, O_A^*) + DI(O_M^*, O_B^*) + DI(O_C^*, O_M^*)$ is lower than the current upper bound then: if O_M is of size n then replace the current upper bound by DI, otherwise go back to step 2.
5. If DI is larger than the current upper bound or O_M is of size n, then stop extending O_M, and consider another possible gene for the last position of O_M, or backtrack to the preceding position if all genes have been considered for the last position.

4 Results

4.1 Branch-and-Bound Efficiency

To measure the efficiency of our branch-and-bound algorithm, we simulated 1,000 DSI histories, each involving i inversions an a unique speciation event, leading to two contemporary genomes (TAGs clusters) of 15 genes and an implicit median containing k genes. Table 1 contains the execution times (on a standard PC) and the average fraction of the search space explored for different values of i and k.

We observe that the execution times depend exponentially on the number of inversions and on the ancestral order size. Nevertheless, it can be used on moderately-sized TAG clusters within reasonable time (43 seconds on average for a history implying an ancestral order of 12 genes and a total of 8 inversions).

4.2 Application on Biological Data

As a first application, we used our branch-and-bound algorithm to infer an ancestral gene order for a pair of human and rat olfactory TAGs clusters. The results are shown in Figure 4. We see that this dataset is compatible with an optimal DLIS history containing only one inversion event, that occurred before the human-rat speciation.

The human cluster is located on chr14@21.2 and the rat cluster on chr15@27.9. Protein sequences and gene orders were obtained from the HORDE database (CLIC #35) [1]. The sequences were aligned with ClustalW [30] and the gene tree generated with MrBayes [14], using the Jones-Taylor-Thornton substitution matrix [19] and performing 1,000,000 MCMC iterations.

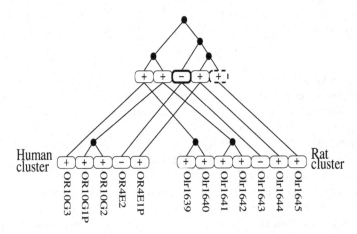

Fig. 4. The ordered reconciled tree obtained for the pair of olfactory receptor TAGs clusters, and the inferred ancestral gene order at the time of human-rat speciation. Transcriptional orientations are indicated by signs. The unique inversion occurred before human-rat speciation and is indicated by a black contour. The rightmost gene in the ancestral TAGs cluster (dashed contour) has its unique descendant in the rat TAGs cluster, indicating a gene loss in the human lineage after the speciation.

This first "simple" application only aims to give an example of a TAGs cluster which is very likely to have evolved in agreement with our model of evolution restricted to simple duplications and inversions, demonstrating its validity.

5   Conclusion

We have presented a formal approach to infer the ancestral gene orders inducing a most parsimonious scenario of inversions in the evolution of a TAGs family in multiple species.

The next step will be to develop heuristic methods to provide good initial solutions for the branch-and-bound algorithm and allow the analysis of larger TAG clusters.

Another important step would be the extension of the model to multiple duplications. However, gene losses are no longer independent from the duplication events in this case [12]. Inferring a tandem duplication tree with multiple duplications and gene losses remains an open problem, even when inversions are not taken into account and only one species is considered.

In addition to our model being restricted to simple duplications, the main problem we face with the inference of TAGs evolutionary histories is the difficulty to obtain a reliable gene tree for some families: events such as gene conversions and unequal crossover can create "mosaic" genes that share more than one ancestor, and pseudogenization is a frequent process. Nevertheless, different strategies could be used to cope with these problems and produce biological knowledge from the present model. For example, the gene tree inference can be facilitated by excluding the pseudogenes of the analysis, and the signal noise can be reduced by choosing closely related species and excluding the period of time that precedes the first speciation from the analysis.

Acknowledgments

This work was supported by grants from the "Fonds Québécois de la Recherche sur la Nature et les Technologies" (D.B. and N.E.M.), the Natural Sciences and Engineering Research Council of Canada (N.E.M.) and the Canadian Institutes of Health Research (M.L.).

References

1. Aloni, R., Olender, T., Lancet, D.: Ancient genomic architecture for mammalian olfactory receptor clusters. Genome Biology 7(10), R88 (2006)
2. Arden, B., Clark, S.P., Kabelitz, D., Mak, T.W.: Human T-cell receptor variable gene segment families. Immunogenetics 42(6), 455–500 (1995)
3. Benson, G., Dong, L.: Reconstructing the duplication history of a tandem repeat. In: ISMB1999. Proceedings of Intelligent Systems in Molecular Biology, Heidelberg, Germany, pp. 44–53. AAAI, Stanford, California, USA (1999)
4. Bergeron, A., Mixtacki, J., Stoye, J.: Reversal distance without hurdles and fortresses. In: Sahinalp, S.C., Muthukrishnan, S.M., Dogrusoz, U. (eds.) CPM 2004. LNCS, vol. 3109, pp. 388–399. Springer, Heidelberg (2004)
5. Bertrand, D., Gascuel, O.: Topological rearrangements and local search method for tandem duplication trees. IEEE Transactions on Computational Biology and Bioinformatics 2(1), 15–28 (2005)
6. Bertrand, D., Lajoie, M., El-Mabrouk, N., Gascuel, O.: Evolution of tandemly repeated sequences through duplication and inversion. In: Bourque, G., El-Mabrouk, N. (eds.) Comparative Genomics. LNCS (LNBI), vol. 4205, pp. 129–140. Springer, Heidelberg (2006)
7. Caprara, A.: The reversal median problem. Journal on Computing 15(1), 93–113 (2003)
8. Chauve, C., Doyon, J.F., El-Mabrouk, N.: Inferring a duplication, speciation and loss history from a gene tree. In: Tesler, G., Durand, D. (eds.) RECOMB 2007. LNCS (LNBI), vol.4751, pp. 45–57. Springer, Heidelberg (2007)
9. Cotton, J.A., Page, R.D.: Going nuclear: gene family evolution and vertebrate phylogeny reconciled. In: Proceedings of the Royal Society B 269, 1555–1561 (2002)
10. Eichler, E., Sankoff, D.: Structural dynamics of eukaryotic chromosome evolution. Science 301, 793–797 (2003)

11. Elemento, O., Gascuel, O.: A fast and accurate distance-based algorithm to reconstruct tandem duplication trees. Bioinformatics 18, 92–99 (2002)
12. Elemento, O., Gascuel, O., Lefranc, M-P.: Reconstructing the duplication history of tandemly repeated genes. Molecular Biology and Evolution 19, 278–288 (2002)
13. Eulenstein, O., Mirkin, B., Vingron, M.: Comparison of annotating duplication, tree mapping, and copying as methods to compare gene trees with species trees. In: Mathematical hierarchies and biology. DIMACS Series in Discrete Mathematics and Theoretical Computer Science, vol. 37 (1997)
14. Huelsenbeck, J.P., Ronquist, F.: MrBayes 3: Bayesian phylogenetic inference under mixed models. Bioinformatics 19(12), 1572–1574 (2003)
15. Fitch, W.M.: Phylogenies constrained by cross-over process as illustrated by human hemoglobins in a thirteen-cycle, eleven amino-acid repeat in human apolipoprotein A-I. Genetics 86, 623–644 (1977)
16. Gascuel, O., Bertrand, D., Elemento, O.: Reconstructing the duplication history of tandemly repeated sequences. In: Gascuel, O. (ed.) Mathematics of Evolution and Phylogeny, pp. 205–235. Oxford University Press, New York (2005)
17. Geraghty, D.E., Koller, B.H., Hansen, J.A., Orr, H.T.: The HLA class I gene family includes at least six genes and twelve pseudogenes and gene fragments. Journal of Immunology 149(6), 1934–1946 (1992)
18. Hannenhalli, S., Pevzner, P.A.: Transforming cabbage into turnip (polynomial algorithm for sorting signed permutations by reversals). Journal of ACM 48, 1–27 (1999)
19. Jones, D., Taylor, W., Thornton, J.: The rapid generation of mutation data matrices from protein sequences. Computer Applications in the Biosciences 8(3), 275–282 (1992)
20. Kaplan, H., Shamir, R., Tarjan, R.E.: A faster and simpler algorithm for sorting signed permutations by reversals. SIAM Journal on Computing 29, 880–892 (2000)
21. Lajoie, M., Bertrand, D., El-Mabrouk, N., Gascuel, O.: Duplication and inversion history of a tandemly repeated genes family. Journal of Computational Biology 14(4), 462–478 (2007)
22. Ma, B., Li, M., Zhang, L.: From gene trees to species trees. SIAM Journal on Computing 30(3), 729–752 (2000)
23. Moret, B., Tang, J., Wang, L., Warnow, T.: Steps toward accurate reconstructions of phylogenies from gene-order data. Journal of Computer and System Science 65(3), 508–525 (2002)
24. Sankoff, D., Blanchette, M.: Multiple genome rearrangement and breakpoint phylogeny. Journal of Computational Biology 5, 555–570 (1998)
25. Shannon, M., Hamilton, A.T., Gordon, L., Branscomb, E., Stubbs, L.: Differential expension of Zinc- Finger transcription factor loci in homologous human and mouse gene clusters. Genome Research 13, 1097–1110 (2003)
26. Shoja, V., Zhang, L.: A roadmap of tandemly arrayed genes in the genomes of human, mouse, and rat. Molecular Biology and Evolution 23(11), 2134–2141 (2006)
27. Tang, M., Waterman, M.S., Yooseph, S.: Zinc finger gene clusters and tandem gene duplication. In: Proceedings of International Conference on Research in Molecular Biology (RECOMB2001), pp. 297–304 (2001)
28. Tesler, G.: GRIMM: genome rearrangements web server. Bioinformatics 18(3), 492–493 (2002)
29. Huntley, S., et al.: A comprehensive catalogue of human krab-associated zinc finger genes: Insights into the evolutionary history of a large family of transcriptional repressors. Genome Research 16, 669–677 (2006)

30. Thompson, J.D., Higgins, D.G., Gibson, T.J.: CLUSTAL W: improving the sensitivity of progressive multiple sequence alignment through sequence weighting, position-specific gap penalties and weight matrix choice. Nucleic Acids Research 22(22), 4673–4680 (1994)
31. Zhang, J., Nei, M.: Evolution of antennapedia-class homeobox genes. Genetics 142(1), 295–303 (1996)
32. Zhang, L., Ma, B., Wang, L., Xu, Y.: Greedy method for inferring tandem duplication history. Bioinformatics 19, 1497–1504 (2003)

Selecting Genomes for Reconstruction of Ancestral Genomes

Guoliang Li[1], Jian Ma[2], and Louxin Zhang[3]

[1] Department of Computer Science
National University of Singapore (NUS), Singapore 117543
ligl@comp.nus.edu.sg
[2] Center for Biomolecular Science and Engineering
University of California at Santa Cruz
Santa Cruz, USA
jianma@soe.ucsc.edu
[3] Department of Mathematics, NUS, Singapore 117543
matzlx@nus.edu.sg

Abstract. It is often impossible to sequence all descendent genomes to reconstruct an ancestral genome. In addition, more genomes do not necessarily give a higher accuracy for the reconstruction of ancestral character states. These facts lead to studying the genome selection for reconstruction problem. In this work, two greedy algorithms for this problem are proposed and tested on computer simulation data as well as a biological example.

1 Introduction

With more and more genomes having been sequenced, reconstructing ancestral proteins and genomic sequences becomes a popular approach for understanding the molecular origins and evolution of key components of virus, bacteria and eukaryotic organisms. Ancestral protein sequences for ribonuclease [8,21], Tu elongation factors [7], and steroid receptors [17] have been reconstructed and validated experimentally. Partial or complete DNA sequences for the common ancestor of placental mammals [1,11], HIV [6], and the 1918 flu virus [15] have also been constructed.

Parsimony, maximum likelihood and Bayesian methods are used for the reconstruction of ancestral protein or DNA sequences (see [4] for details of these methods). The reconstruction accuracy of these methods has been assessed by both theoretical analysis [13,19,14] and random simulation [20,1,2,18]. These analyses indicate that the topology of the phylogenetic tree relating the extant genomes to the target ancestral genomes affects the reconstruction accuracy significantly. For example, a starlike phylogeny allows the ancestral character states to be more accurately inferred than other topologies [14,3] although the actual situation is much more complicated [10]. Intuitively, more genomes should give better reconstruction accuracy at the root of a phylogeny. However, this is not always true even for a simple method like parsimony. Recently, we have shown

G. Tesler and D. Durand (Eds.): RECOMB-CG 2007, LNBI 4751, pp. 110–121, 2007.

that, in many phylogenetic trees, the accuracy of the ancestral state in the root reconstructed with all the genomes is smaller than the accuracy of the ancestral state reconstructed with only one genome (Refer to Section 3, and also see [9] for details). This motivates us to study the following computational problem:

> Given a phylogeny P on a set of genomes, an integer k and a reconstruction method $\mathcal{M}$, find a subset of k genomes in the phylogeny that gives the highest accuracy of reconstructing the ancestral genome at the root of the phylogeny, using method $\mathcal{M}$.

Another motivation for studying this problem is that, due to resource constraint, it is often impossible to sequence all the extant genomes that are evolved from the target ancestral genome. In this paper, we study the above genome selection for reconstruction problem. We develop two greedy algorithms for it and test them with the Fitch method on random simulation data as well as a biological example.

The rest of this paper is divided into six sections. In Section 2, we briefly introduce the Fitch method and its accuracy analysis in a simple Jukes-Cantor model. In Section 3, we demonstrate that more genomes are not necessarily better in accuracy for reconstructing an ancestral genome. In Section 4, we present two greedy algorithms for the genome selection for reconstruction problem. In Section 5, we test our algorithms against random phylogenetic trees. In Section 6, we examine a biological example. In Section 7, we conclude the paper with a few of remarks.

2   Parsimony Methods and Its Accuracy

2.1   A Simple Jukes-Cantor Evolutionary Model

Given the phylogenetic tree for a group of species, we assume that the character evolves by a Markov process, starting with a state at the root and proceeding to the leaves node by node. The probability that a node x receives a state t_x depends only on its parent node p and the conditions along the branch from p to x. The evolutionary model specifies the probability that a character c evolves to a character d on a branch from p to x as a conditional probability $\Pr[s_x = d | s_p = c]$. Here, we consider a simple Jukes-Cantor model. In this symmetric model, there are only two states, say 0 and 1, and the probability of a substitution change of any sort on any branch would be the same.

2.2   Parsimony Reconstruction Method

For reconstructing character evolution, parsimony methods assign to each internal node those states that allow for the fewest number of substitutions throughout the tree. In this paper, we study the genome selection for reconstruction problem with respect to the parsimony method proposed by Fitch [5]. This parsimony method assigns a set of states to each node one by one downward through the tree, starting with the leaves and using the subsets previously computed for the node's children. For each leaf node, the observed state forms the state set. Assume A is an internal node with children B and C. The following rule is used to compute the state subset S_A from the state subsets S_B and S_C:

$$S_A = \begin{cases} S_B \cup S_C & if \ S_B \cap S_C = \phi, \\ S_B \cap S_C & if \ S_B \cap S_C \neq \phi. \end{cases}$$

The state set at the root contains all the possible states that will be assigned to it. We say that the method unambiguously reconstructs a state at the root if the state set contains only that state and ambiguously reconstructs a state if the state set contains both 0 and 1.

Note that the method presented in [13] (see also [12]) reconstructs the states of the internal nodes based on the information from all the leaf nodes, which is a little bit more complicated than the method described above. As far as the accuracy of the root is concerned, it gives the same state set as the method described above and hence has the same reconstruction accuracy at the root.

2.3   Reconstruction Accuracy

Assume the character evolves in a phylogeny with the root A according to a probabilistic evolutionary model. The evolutionary model specifies a prior probability for each state at A. When we say D is a state configuration at the leaves, we mean that it contains a state for each leaf in the phylogenetic tree. For a state c and a state configuration D at the leaves, we let $P[D|c]$ to denote the probability that the state c at the root evolves into the states given by D at the leaves in the phylogeny. Then the reconstruction accuracy of a method M is

$$P_{accuracy} = \sum_{c,D} prior(c)P[D|c]I(c, D, M),$$

where $I(c, D, M) = 1$ if the method M reconstructs c correctly from D at the root and 0 otherwise.

In this paper, we consider a symmetric evolutionary mode with two states 0 and 1. Hence, the reconstruction accuracy is independent of the prior distribution of the states. The unambiguous reconstruction accuracy of the Fitch method is

$$P_{accuracy} = \sum_{D} P[D|0]I(0, D, M) = \sum_{D} P[D|1]I(1, D, M).$$

There are three different state subsets $\{0\}$, $\{1\}$ and $\{0, 1\}$ with two states 0 and 1. For a state set t, and a state s, we use $P_N[t|s]$ to denote the probability that the state set t is computed at the node N by the Fitch's method given the true state s at N. It is not hard to see that $P_{accuracy} = P_A[\{0\}|0] = P_A[\{1\}|1]$.

At a leaf x with observed state s, we have

$$P_x[\{s\}|s] = 1, \ P_x[\{s'\}|s] = P_x[\{0, 1\}|s] = 0$$

for $s' \neq s$. Let N be an internal node with the children L and R. Then, for $c, d = 0, 1$,

$$
\begin{aligned}
& P_N[\{d\}|c] \\
& = \sum_{x,y=0,1} \Pr[s_L = x|s_N = c] \Pr[s_R = y|s_N = c] P_L[\{d\}|x] P_R[\{d\}|y] \\
& + \sum_{x,y=0,1} \Pr[s_L = x|s_N = c] \Pr[s_R = y|s_N = c] \\
& \times \{ P_L[\{d\}|x] P_R[\{0,1\}|y] + P_L[\{0,1\}|x] P_R[\{d\}|y] \}
\end{aligned}
$$

and

$$
P_N[\{0,1\}|c] = 1 - P_N[\{0\}|c] - P_N[\{1\}|c].
$$

The above recurrence relations give immediately a dynamic programming approach for computing the reconstruction accuracy of the Fitch's method, which is used in our analysis in the rest of this paper. Such a method first appeared in [13].

The ambiguous reconstruction accuracy of the method takes the ambiguous state into consideration and is defined as

$$
P_{\text{A-accuracy}} = P_A[\{1\}|1] + \frac{1}{2} P_A[\{0,1\}|1]
$$

where the first term is the unambiguous reconstruction accuracy and the second term in the expression simply says that, when either state 0 or 1 is equally parsimonious as a root state, we select either state with equal probability.

3   More Genomes Are Not Necessarily Better

Counterintuitively, more genomes do not necessarily give better reconstruction even for the parsimony methods [9]. The reason is that the reconstruction accuracy is highly sensitive to the topology used for the reconstruction and more genomes may introduce more noise in the reconstructed ancestral state. For completeness, we briefly summarize the partial results proved in [9] in this section.

We first consider the complete phylogenetic trees. Let T be the complete phylogeny with 4 leaves shown in Figure 1(a). We assume the conservation probability is p on any branch in T and $q = 1 - p$. For each node N, we denote the true state at N by s_N. Then, the conservation probability on each path from the root to a leaf is

$$
\begin{aligned}
& P_{path} \\
& = \Pr[s_x = 1|s_A = 1] \\
& = \Pr[s_B = 1|s_A = 1] \Pr[s_x = 1|s_B = 1] + \Pr[s_B = 0|s_A = 1] \\
& \quad \times \Pr[s_x = 1|s_B = 0] \\
& = p^2 + q^2,
\end{aligned}
$$

where we assume x is a leaf below the node B.

Let t_N be the reconstructed state set at a node N. For $V = B, C$ and $s, s' = 0, 1$,

$$
\Pr[t_V|V = s] = \begin{cases} q^2 & \text{if } t_V = \{s'\} \text{ and } s' \neq s, \\ p^2 & \text{if } t_V = \{s'\} \text{ and } s' = s, \\ 2pq & \text{if } t_V = \{0,1\}. \end{cases}
$$

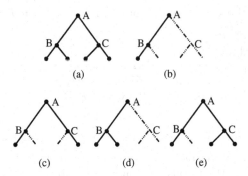

Fig. 1. (a) The complete phylogeny with 4 leaves; (b) The path topology; (c) The $\wedge$-shape topology; (d) The Y-shape topology; (e) The W-shape topology

By this formula, the unambiguous reconstruction accuracy of using all the four taxa is

$$
\begin{aligned}
&P_{whole} \\
&= \Pr[t_A = \{1\}|s_A = 1] \\
&= \sum_{x,y\in\{0,1\}} \Pr[s_B = x|s_A = 1]\Pr[s_C = y|s_A = 1]\Pr[t_B = \{1\}|s_B = x] \\
&\quad \times \Pr[t_C = \{1\}|s_C = y] \\
&\quad + \sum_{x,y\in\{0,1\}} \Pr[s_B = x|s_A = 1]\Pr[s_C = y|s_A = 1] \\
&\quad \times \{\Pr[t_B = \{1\}|s_B = x]\Pr[t_C = \{0,1\}|s_C = y] \\
&\quad + \Pr[t_B = \{0,1\}|s_B = x]\Pr[t_C = \{1\}|s_C = y]\} \\
&= (p^6 + q^6 + 2p^3q^3) + 2[2p^5q + 2pq^5 + 2p^2q^4 + 2p^4q^2] \\
&= (p^3 + q^3)^2 + 4[p^2q(p^3 + q^3) + pq^2(p^3 + q^3)] \\
&= (p^2 + q^2 - pq)(1 + pq).
\end{aligned}
$$

Since

$$
\begin{aligned}
&P_{path} - P_{whole} \\
&= (p^2 + q^2) - (p^2 + q^2 - pq)(1 + pq) \\
&= -(p^2 + q^2)pq + pq(1 + pq) \\
&= 3p^2q^2.
\end{aligned}
$$

we have $P_{path} > P_{whole}$ unless $p = 0, 1$. Similarly, we can also show that the unambiguous reconstruction accuracy of using the topologies in Figure 1(c), 1(d) and 1(e) is smaller than P_{path}.

To find out how often the accuracy of using the whole phylogenetic tree to reconstruct ancestral character states at the root is smaller than the conservation probability on a path from the root to a leaf, we conducted simulation test by generating random phylogenetic trees in the Yule model.

In the Yule model, a random phylogenetic tree grows recursively from a single root node. In each step, one leaf in the current tree is selected to add two children with equal probability. The procedure repeats until a phylogenetic tree with the required number of leaves is generated.

For each set (N, p) of parameters, we generated five thousand random phylogenetic trees and count how many trees have the reconstruction accuracy less

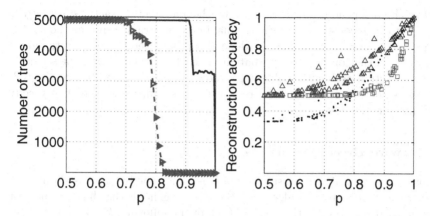

Fig. 2. (a) The number of the random phylogenetic trees in which the unambiguous reconstruction accuracy from the longest (triangle line) or shortest (dot line) path is better than the unambiguous reconstruction accuracy from the whole phylogeny. (b) The unambiguous reconstruction accuracy of the whole tree (dot curve), shortest path (triangle curve) and longest path (square curve).

than the conservation probability on the shortest or longest path from the root to a leaf. Here, N denotes the number of leaves in the random trees and is set to nine, fifteen, or twenty; p represents the conservation probability on each branch and is set to $0.5 + 0.01i$ for each $0 \leq i \leq 49$.

Figure 2(a) shows the number of randomly-generated phylogenetic trees in which the conservation probability on the shortest path or longest path is larger than the reconstruction accuracy of using the whole tree. When the conservation probability p on each branch is in the range of 0.5 and 0.8, the conservation probability on the shortest path to some leaf is better in most of the trees. When p exceeds 0.9, the number of 'bad' trees decreases rapidly. Figure 2(b) shows sampled reconstruction accuracy of these three different reconstructions.

We show the fact the more genomes do not always give better unambiguous reconstruction accuracy in some phylogenetic trees. This observed fact also holds for the ambiguous reconstruction accuracy [9].

4   Algorithms for Genome Selection

The counterintuitive observation in above section and the fact that limited resource prohibits one to sequence all the descendent genomes for ancestral reconstruction motivate us to study the genome selection for reconstruction problem. Formally, this problem is defined as

Genome selection for reconstruction
Instance: A phylogenetic tree P on a set of n genomes, a number k and a reconstruction method $\mathcal{M}$.

Question: Find k genomes in P that allows the ancestral character states at the root of P to be reconstructed with the maximum accuracy, using method $\mathcal{M}$.

Since the reconstruction accuracy depends on both the topology of the given phylogeny and the conservation probability of each branch, the genome selection for reconstruction problem is unlikely polynomial-time solvable although its NP-hardness is not proved yet. In the rest of this section, we present two greedy algorithms for it.

4.1   Forward Greedy Algorithm

The forward greedy algorithm selects the k genomes one by one based on accuracy increment. Initially, the algorithm chooses the genome that has the shortest evolutionary distance from the root. In each of the following $k - 1$ steps, the algorithm selects a genome that gives the maximum increment on reconstruction accuracy. In summary, the forward greedy algorithm can be described as follows:

FORWARD GREEDY ALGORITHM

```
1. Set S ← φ;
2. Add the nearest genome to S;
3. For i = 1, 2, ···, k − 1 do {
       for each genome g not in S, compute the accuracy A_g
         of the reconstruction by applying M to S ∪ {g};
       Add g to S if A_g is the maximum over all gs;
       }
4. Output S.
```

4.2   Backward Greedy Algorithm

The backward greedy algorithm removes $n-k$ genomes one by one by considering the accuracy decrease. Initially, there are n genomes. In each of $n - k$ steps, the algorithm selects a genome whose removal leads to the least decrease in reconstruction accuracy.

BACKWARD GREEDY ALGORITHM

```
1. Let S contain all the genomes in the phylogeny;
2. For i = 1, 2, ···, n − k do {
       for each genome g in S, compute the accuracy A_g
         of the reconstruction by applying M to S − {g};
       Remove g from S if A_g is the maximum over all g's;
       }
3. Ouput S.
```

Since the backward greedy algorithm starts from the full phylogeny, it is not hard to see that the backward greedy algorithm is not efficient as the forward greedy algorithm especially when reconstruction method such as the maximum likelihood, is used. However, as we will see below, this method has better performance.

5 Simulation Test

To evaluate the performance of the forward and backward greedy algorithms, we apply them with the Fitch method on random phylogenetic trees generated in the Yule model. For $p = 0.75, 0.80, 0.85, 0.90, 0.95, 0.99$ and $N = 9, 16$, we respectively generated one hundred balanced and one hundred imbalanced random trees with N leaves using the method described in the previous section.

For each random tree with nine leaves, we apply the two greedy algorithms to find a three-leaf subset and a six-leaf subset; for each random tree with sixteen leaves, we apply the two greedy algorithms to find a five-leaf subset and a ten-leaf subset. The accuracy of reconstructing the character states at the root using the found subset is computed and compared with the optimal accuracy over all the subsets containing the desired number of genomes and the accuracy of using all the genomes. Figure 3 shows the average accuracies from different algorithms on one hundred balanced random trees. The left bar graph is the average accuracy of six-leaf subsets from the balanced random phylogeny with nine leaf nodes, and the right bar graph is the average accuracy of ten-leaf subsets from the balanced random phylogeny with sixteen leaf nodes. The performance of the greedy algorithms on the imbalanced trees with $p < 0.9$ is generally better (data not shown here due to space limitation), which is consistent with the results in Section 3.

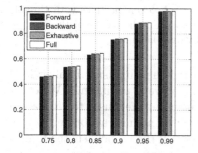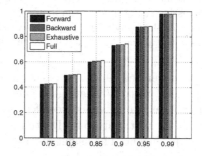

Fig. 3. The left and right bar graphs summarize the average reconstruction accuracy of the subsets found by the two algorithms against the optimal accuracy on the randomly generated phylogenetic trees with nine and sixteen leaves respectively

In these tests, the algorithms can identify subsets of genomes that result in better reconstruction accuracy than obtained using all genomes in the tree. When the conservation probability $p = 0.75$, the accuracy from the backward greedy algorithm on the three-leaf subset of the nine-leaf phylogeny is always better than the accuracy from the full phylogeny (data not shown). As the conservation probability increases, the greedy algorithms obtain better accuracy less frequently.

Both tests also indicate that the backward greedy algorithm yields higher reconstruction accuracy than the forward greedy algorithm in about 80% random trees. But, as we mentioned earlier, the drawback of the backward greedy algorithm is that it is time-consuming, especially when the phylogenetic tree is large and maximum likelihood or a Bayesian method is used for reconstruction.

Furthermore, Figure 3 shows that, on average, the accuracy from the greedy algorithms are comparable to, if not better than, the accuracy from the full phylogeny. This provides the support for selecting a subset of the genomes to reconstruct the ancestral genomes when there are resource constraints and we can not sequence all the extant genomes in the domain of interest.

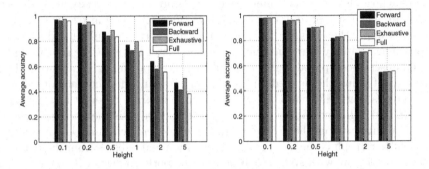

Fig. 4. Average accuracy under different tree heights. The left and right bar graphs summarize the average unambiguous reconstruction accuracy and ambiguous reconstruction of the subsets found by the two algorithms against the optimal accuracy from exhaustive search on the randomly generated phylogenetic trees with three-leaf subsets on nine-leaf phylogeny respectively. The x-axis is the height of the trees and the y-axis is the average accuracy under the individual heights.

We also generated random trees with different heights using Evolver in the PAML package (http://abacus.gene.ucl.ac.uk/software/paml.html). We considered the trees with nine and sixteen leaves. The parameters used to generate the trees are: 10 for Birth rate, 5 for Death rate, 1 for Sampling fraction, and 0.1, 0.2, 0.5, 1, 2, 5 for height. The height means the sum of the branch lengths from the root to all leaf nodes. For each possible conbination of parameter values, we generated one hundred random trees and estimated the transition probability along each branch using the Jukes-Cantor model. The left panel of Figure 4 shows the average unambiguous reconstruction accuracy for different heights. The right panel of Figure 4 shows the average ambiguous reconstruction accuracy for different heights. In both cases, the solutions output by our greedy algorithms are near optimal. Note also that, for ambiguous reconstruction accuracy, the backward greedy algorithm outperforms the forward greedy algorithm.

Unlike the unambiguous reconstruction accuracy, the ambiguous reconstruction accuracy from the full phylogeny is better on average, indicating that more genomes introduce more noise. It seems true that more genomes always result in higher ambiguous reconstruction accuracy in an ultrametric phylogenetic tree

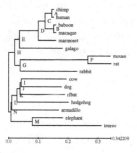
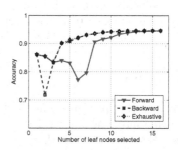

Fig. 5. A phylogenetic tree in the reconstruction of the Boreoeutherian ancestor and the unambiguous reconstruction accuracies of the greedy algorithms on this tree. In the left graph, the branch lengths are the substitution rate. In the right graph, the solid line with triangles represents the accuracies from the forward greedy algorithm, the dashed line with squares represents the accuracies from the backward greedy algorithm, and the dotted line with diamonds represents the accuracies from the exhaustive search.

in which all the paths to a leaf have the equal height. However, it is not known how to prove this hypothesis.

6   A Biological Example

We consider the reconstruction of the so-called Boreoeutherian ancestor where a rapid radiation of many different lineages occurred. We applied the both forward and backward algorithms to the phylogeny shown in Figure 5 (see [16]). Among the sixteen extant species of this phylogenetic tree, the genomes of human, chimp, macaque, rat, mouse, and dog have been sequenced for the first time; other genomes have been partially sequenced.

In this example, we examined the expected accuracy of the reconstruction of the four nucleotides on each base in the Boreoeutherian ancestor. The branch weight in the phylogeny is the substitution rate. Therefore, under the Jukes-Cantor model, we assume that, for each branch, the conservation probability is one minus the branch weight and the probability of one nucleotide replacing another is one third of the substitution rate. Since the true ancestral nucleotide residues are unknown at the Boreoeutherian ancestor, it is impossible to obtain the true reconstruction accuracy. As a result, we calculated the expected reconstruction accuracy using the formula stated in Section 2.3. (Here, we considered four states, rather than two states in the model used in section 2 and section 3.)

For each of k from one to sixteen, the reconstruction accuracy obtained using k genomes, estimated by the greedy algorithms, are compared in Figure 5. When $k = 1, 2$, the forward greedy algorithm performed similarly to the exhaustive search algorithm. When $k = 3$, all three algorithms obtained similar accuracy. When $k > 3$, the performance of the backward greedy algorithm is similar to the exhaustive search algorithm, and the performance of the forward greedy algorithm is worse. For example, when $k = 8$, the backward algorithm output the following genomes: human, dog, galago, mouse, rabbit, dog, armadillo, elephant, leading to the unambiguous reconstruction accuracy as high as 93.6%, which is quite near the accuracy 94.6% obtained using the full phylogeny.

7   Conclusion

It is well known that parsimony method is not consistent when the branches are long more characters do not lead to the right phylogeny (see Chapter 9 of [4] for details). Here, we observe that more genomes are not necessarily better in the reconstruction of ancestral character states with a given phylogeny, giving a complementary example in which more data is not necessarily better.

Motivated by the above counterintuitive result and the impossibility of sequencing all the descendent genomes for ancestral genome reconstruction, we have studied the genome selection for reconstruction problem in this work. We proposed two greedy algorithms for the problem and tested them with simulation data. The experiment results showed that, in most of the cases, the accuracy from the greedy algorithms is comparable to the highest accuracy of using the same number of genomes; it is also comparable to, if not better than, the accuracy of using all the genomes in the full phylogeny. In general, the forward algorithm is more straightforward, but has poor performance compared with the backward greedy algorithm.

We also tested our algorithms on the reconstruction of the Boreoeutherian ancestor of the placental mammals. The test shows that using only eight genomes identified by the backward greedy algorithm, an expected reconstruction accuracy of 93.6% can be obtained. It is quite close to the accuracy obtained with the full phylogeny, namely 94.6%. This indicates that selecting the genomes for ancestral genome reconstruction is also practical.

Acknowledgment

The authors would like to thank the reviewers and D. Durand for their valuable suggestions on revising the paper. LX Zhang gratefully acknowledged the NUS ARF grant R-146-000-068-112 and NSFChina3052802 for partially supporting this project. He also thanks Webb Miller for stimulating this research by pointing out the paper [10] to him.

References

1. Blanchette, M., Green, E.D., Miller, W., Haussler, D.: Reconstructing large regions of an ancestral mammalian genome in silico. Genome Res. 14, 2412–2423 (2004)
2. Cai, W., Pei, J.M., Grishin, N.V.: Reconstruction of ancestral protein sequences and its application. BMC Evol. Biol. 4, e33 (2004)
3. Evens, W., Kenyon, C., Peres, Y., Schulman, L.J.: Broadcasting on trees and the ising model. Annals of Applied Prob. 10, 410–433 (2000)

4. Felsenstein, J.: Inferring Phylogenies, Sinauer Associates. Sunderland, Massachusetts (2004)
5. Fitch, W.M.: Toward Defining the Course of Evolution: Minimum Change for a Specific Tree Topology. Systematic Zoology 20, 406–416 (1971)
6. Hillis, D.M., Huelsenbeck, J.P., Cunningham, C.W.: Application and accuracy of molecular phylogenies. Science 264, 671–677 (1994)
7. Gaucher, E.A., Thomson, J.M., Burgan, M.F., Benner, S.A.: Inferring the palaeoenvironment of ancient bacteria on the basis of resurrected proteins. Nature, 285–288 (2003)
8. Jermann, T.M., Opitz, J.G., Stackhouse, J., Benner, S.A.: Reconstructing the evolutionary history of the artiodactyl ribonuclease superfamily. Nature 374, 57–59 (1995)
9. Li, G.L., Steel, M., Zhang, L.X.: More taxa are not necessarily better for the reconstruction of ancestral sequence by parsimony. Manuscript (2007)
10. Lucena, B., Haussler, D.: Counterexample to a claim about the reconstruction of ancestral character states. Syst. Biol. 54, 693–695 (2005)
11. Ma, J., Zhang, L., Suh, B.B., Raney, B.J., Burhans, R.C., Kent, W.J., Blanchette, M., Haussler, D., Miller, W.: Reconstructing contiguous regions of an ancestral genome. Genome Res. 16, 1557–1565 (2006)
12. Maddison, W.P., Maddison, D.R.: MacClade: analysis of phylogeny and character evolution. Version 3, Sinauer, Sunderland, MA
13. Maddison, W.P.: Calculating the probability distributions of ancestral states reconstructed by parsimony on phylogenetic trees. Systematic Biology 44, 474–481 (1995)
14. Schultz, T.R., Cocroft, R.B., Churchill, G.A.: he reconstruction of ancestral character states. Evolution 50, 504–511 (1996)
15. Taubenberger, J.K., Reid, A.H., Lourens, R.M., Wang, R., Jin, G., Fanning, T.: Characterization of the 1918 influenza virus polymerase genes. Nature 437, 889–893 (2005)
16. The ENCODE Project Consortium, Identification and analysis of functional elements in 1% of the human genome by the ENCODE pilot project, Nature, 447, 799–816 (2007)
17. Thornton, J.W., Need, E., Crews, D.: Resurrecting the ancestral steroid receptor: ancient origin of estrogen signaling. Science 301, 1714–1717 (2003)
18. Williams, P.D., Pollock, D.D., Blackburne, B.P., Goldstein, R.A.: Assessing the accuracy of ancestral protein reconstruction methods. PLoS Comput Biol. 2, e69 (2006)
19. Yang, Z.H., Kumar, S., Nei, M.: A new method of inference of ancestral nucleotide and amino acid sequences. Genetics 141, 1641–1650 (1995)
20. Zhang, J., Nei, M.: Accuracies of ancestral amino acid sequences inferred by parsimony, likelihood, and distance methods. J. Mol. Evol. 44(S1), 139–146 (1997)
21. Zhang, J., Rosenberg, H.F.: Complementary advantageous substitutions in the evolution of an antiviral RNase of higher primates. Proc. Natl. Acad. Sci. USA 99, 5486–5491 (2002)

A Heuristic Algorithm for Reconstructing Ancestral Gene Orders with Duplications

Jian Ma[1], Aakrosh Ratan[2], Louxin Zhang[3], Webb Miller[2], and David Haussler[1]

[1] Center for Biomolecular Science and Engineering,
University of California, Santa Cruz, CA 95064, USA
`jianma@soe.ucsc.edu`
[2] Center for Comparative Genomics and Bioinformatics
Penn State University, University Park, PA 16802, USA
[3] Department of Mathematics,
National University of Singapore, Singapore 117543

Abstract. Accurately reconstructing the large-scale gene order in an ancestral genome is a critical step to better understand genome evolution. In this paper, we propose a heuristic algorithm for reconstructing ancestral genomic orders with duplications. The method starts from the order of genes in modern genomes and predicts predecessor and successor relationships in the ancestor. Then a greedy algorithm is used to reconstruct the ancestral orders by connecting genes into contiguous regions based on predicted adjacencies. Computer simulation was used to validate the algorithm. We also applied the method to reconstruct the ancestral genomes of ciliate *Paramecium tetraurelia*.

Keywords: gene order reconstruction, duplication, contiguous ancestral region.

1 Introduction

The increasing number of genome sequences becoming available makes it feasible to computationally reconstruct ancient genomes of related species that have undergone genome rearrangements. The heart of this problem is to "undo" these large scale rearrangements and restore the ancestral gene order. Previous studies mainly focused on solving the median problem, which is either based on reversal (inversion) distance or breakpoint distance. In this problem one tries to reconstruct the common ancestor of two descendant genomes using one additional outgroup genome. Unfortunately, the median problem doesn't have exact and efficient algorithms [1,2]. In the past, heuristic programs for both breakpoint median problem and reversal median problem have been proposed [3,4,5]. But the discrepancy between the computational prediction and the result from cytogenetic experiments [6,7] suggests a need to explore further computational methods for ancestral genome reconstruction.

In our recent work [8], we proposed a new approach for reconstructing the ancestral order based on the adjacencies of orthologous genomic content in modern

G. Tesler and D. Durand (Eds.): RECOMB-CG 2007, LNBI 4751, pp. 122–135, 2007.

species, which essentially avoids solving any rearrangement median problem. The critical procedure of the method is analogous to Fitch's parsimony algorithm [9]. Instead of inferring ancestral nucleotides, we infer the locally parsimonious predecessor and successor relationships of the orthologous conserved segments in the ancestor, in this case the ancestor of most placental mammals, known as the Boreoeutherian ancestor. Another procedure then connects these segments into 29 contiguous ancestral regions (CARs). Our result agrees with the cytogenetic prediction fairly well [10].

However, the main drawback of the method in [8] is that it doesn't handle duplications. Indeed, duplications (including segmental duplications and tandem duplications) have a great impact on genome evolution [11]. Some previous theoretic studies [12,13,14] have included duplications (sometimes with loss) along with rearrangements. In this paper, we extend the method in [8] and propose an efficient heuristic approach to incorporate duplications into analysis when we are inferring ancestral gene orders.

2 Methods

2.1 Definitions

In this paper, we use the term **gene** to represent an atomic evolutionary unit that has never been broken due to breakpoints caused by any operations (duplication or rearrangement). If two genes are derived from a common ancestral gene, then they belong to the same **gene family**. We use $g[x]$ to represent the gene x in genome g. Also, if two genes from the same family x are in the same genome g, then we denote these genes as $g[x.i]$ and $g[x.j]$ ($i \neq j$). A **chromosome** of a modern or ancestral genome consists of a list of genes where each gene has a sign (orientation) that is either positive (+) or negative (−). The **reverse complement** of a chromosome is obtained by reversing the list and flipping the sign of each gene. A **genome** is a set of chromosomes.

If genome g contains gene x, then the **predecessor** $p_g(x)$ is defined as the gene that immediately precedes x on the same chromosome. Predecessor has a sign. In the opposite orientation, $p_g(-x)$ immediately precedes $-x$ in the reverse complement of the same chromosome. We set $p_g(x) = \Phi_A$ if x appears first on a chromosome. The **successor** $s_g(x)$ of x is defined analogously. And we also set $s_g(x) = \Phi_Z$ if x appears last on a chromosome. For instance, let g have the chromosome (1 −4.1 −3 4.2 5 2). Then $p_g(1) = \Phi_A$, $p_g(2) = 5$, $p_g(-3) = -4.1$, $s_g(-4.1) = -3$, $p_g(-1) = 4.1$, $s_g(-5) = -4.2$, etc.

In addition to speciation events, the original ancestral genes evolve through large-scale evolutionary operations which include insertion/deletion, rearrangements (inversion, translocation, fusion/fission), and tandem and segmental duplications. Consequently, we have a different number of genes and different gene orders in present day genomes. Our goal is to reconstruct the order and orientation of genes in the target ancestral genome. We call each reconstructed chromosome a **contiguous ancestral region (CAR)**.

2.2   Species Tree, Gene Tree, and Reconciled Tree

A **species tree** is a full binary tree describing the phylogeny among different species (Fig.1(A)). All the bifurcating ancestral nodes represent speciation events, while leaves correspond to modern species. Each branch in the tree has branch length d indicating the evolutionary distance. Along the branch between two species (from ancestor to descendant), evolutionary operations could happen. In this paper, we assume that the species tree is already known, and it has been rooted and directed.

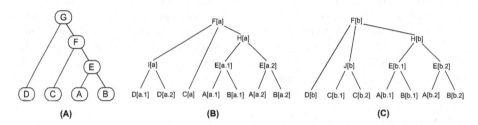

Fig. 1. (A) Species tree of modern species A, B, C, and D. Gene trees of gene family a and b are in (B) and (C), respectively. Branch length $d(D[a.1], I[a]) + d(I[a], F[a])$ in (B) is equivalent to the branch length $d(D, G) + d(F, G)$ in the species tree. We also have $d(D[b], F[b]) > d(D, G) + d(F, G)$. For other branch lengths in the gene trees, we have: $d(A[a.1], E[a.1]) = d(A[a.2], E[a.2]) = d(A[b.1], E[b.1]) = d(A[b.2], E[b.2])$, $d(B[a.1], E[a.1]) = d(B[a.2], E[a.2]) = d(B[b.1], E[b.1]) = d(B[b.2], E[b.2])$, $d(D[a.1], I[a]) = d(D[a.2], I[a])$, $d(C[b.1], J[b]) = d(C[b.2], J[b])$, $d(E[a.1], H[a]) = d(E[a.2], H[a]) = d(E[b.1], H[b]) = d(E[b.2], H[b])$, $d(H[a], F[a]) = d(H[b], F[b])$, $d(C[a], F[a]) = d(C[b.1], J[b]) + d(J[b], F[b])$.

A **gene tree**, on the other hand, is an unrooted tree, characterizing the relationships among genes in the same gene family across different species (Fig.1(B) and (C)). It also has branch lengths associated with each branch in the tree. In this paper, we have two assumptions for gene trees: (1) the duplication events have been dated and they are consistent with what happened in nature, e.g. duplication event $I[a]$ in Fig.1(B); (2) in the gene tree, all the branch lengths are exact. Therefore, if in the following reconciliation step, a node in the gene tree turns out to correspond to a speciation event, then it has a perfect match to the node in the species tree, e.g. in Fig.1 the distance from $A[a.1]$ to $E[a.1]$ in (B) is exactly the same as the distance from A to E in (A), i.e. $d(A[a.1], E[a.1]) = d(A, E)$.

A **reconciled tree** is a mapping between all gene trees and the species tree with gene duplications and losses being postulated [15]. In order to get the reconciled tree, we merge the unrooted gene trees into the rooted species tree. A reconciled tree, denoted as $\mathbb{T}$, represents all speciation and duplication events that have left a record of their effects in the leaf genomes. We start with the species tree and reconcile the gene trees into it one at a time. The species tree as well as the two gene trees in Fig.1 can be reconciled into Fig.2(A). Our reconciliation algorithm is less complicated than the traditional methods, e.g. [16] and [17], because in our case the true species tree is known and the distances in the gene trees are exact. (See Appendix for detailed reconciliation algorithm).

Each reconciliation labels the bifurcating nodes of the gene tree being reconciled as either duplication nodes or speciation nodes, maps the speciation nodes to the corresponding speciation nodes in the species tree, and maps the duplication nodes to inferred duplication nodes along the branches of the species tree (Fig.2(A)). The final reconciled tree includes these additional duplication nodes (Fig.2(B)). Each node in the reconciled tree is a genome. If there are duplications that occurred before the root of the species tree, then the root of the reconciled tree is an ancestor of the species tree root, and these ancient duplications are represented on an additional path leading from the root in reconciled tree to the root of the species tree within it, e.g. node K in Fig.2.

During the reconciliation, genes are also added along the branches of each gene tree to represent intermediate forms that are inferred to have existed at duplication branches but do not appear in the original gene tree for the family (Fig.2(A), e.g. gene a in J). The resulting gene trees are called **augmented gene trees** and denoted T_a for gene family a. For each node x in T_a, there is a **mapping** ϕ that maps x to a node y in $\mathbb{T}$, i.e. $y = \phi(x)$, indicating the genome y that gene x belongs to. Also, the root of an augmented gene tree need not always

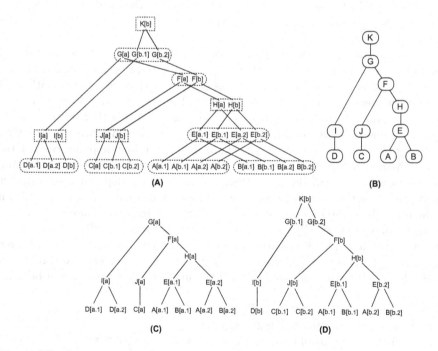

Fig. 2. (A) The reconciled tree from species tree and gene trees of gene family a and b. Node I, J, K, and H show four duplications; K is an ancient duplication. (B) A simplified form of reconciled tree $\mathbb{T}$. (C) augmented gene tree T_a. (D) T_b.

map to the root of the reconciled tree. If the gene family is first introduced by an insertion event, then the last common ancestor in the reconciled tree of all the observed genes in the family may be a node below the root. For example, in T_a the root does not map to K in the reconciled tree. We could interpret gene family a as an insertion before G but after K in the reconciled tree.

Along branch h to g in the reconciled tree, we define $\tilde{A}_h(g[x])$ as the **direct ancestor** of $g[x]$ in h and $\tilde{D}_g(h[x])$ as the set of **direct descendants** of $h[x]$ in g. Note that $\tilde{D}(h[x])$ could contain two descendants if x is duplicated at h. If $g[x]$ has no ancestor, then $\tilde{A}_h(g[x]) = \emptyset$. Conversely, if $h[x]$ has no descendant in g, $\tilde{D}_g(h[x]) = \emptyset$. For example, in Fig.2, $\tilde{A}_E(B[a.1]) = E[a.1]$, $\tilde{D}_C(J[b]) = \{C[b.1], C[b.2]\}$.

2.3   Reconstructing Ancestral Adjacency

After obtaining a reconciled tree $\mathbb{T}$ and augmented gene trees T_i (for all gene family i), our goal is to determine a set of lists of gene orders that closely approximates the genome structure of the species corresponding to a target ancestral genome in $\mathbb{T}$.

For any genome g, we associate with each gene x two sets of signed genes, denoted $P_g(x)$ and $S_g(x)$, giving potential predecessors and successors of x relative to chromosomes of g. If g is a modern genome, $P_g(x) = \{p_g(x)\}$ and $S_g(x) = \{s_g(x)\}$, for each x. If g does not contain x, then both sets are empty. We also define that $\tilde{A}_h(P_g(x)) = \{\tilde{A}_h(y_i) \mid y_i \in P_g(x)\}$. $\tilde{D}_h(P_g(x))$ can be defined analogously.

We use N_g to denote the number of genes in genome g, which can be counted directly from the reconciled tree. For example, $N_E = 4$ in the example in Fig.2.

The inference procedures of predecessor and successor associated with each gene in the gene tree is similar to the method in [8]. The first stage of the algorithm works in a bottom-up fashion. The general idea is that, for each node π in the gene tree, we compute its predecessor set according to the following rule: If π is a leaf, then predecessor set consists of the unique predecessor. Otherwise, assume π has children τ and φ; then, the predecessor set is equal to the intersection or union of the predecessor sets of τ and φ depending on whether their predecessor sets are disjoint or not. The second stage works in a top-down fashion to adjust the predecessor sets. Similarly, we infer the successors.

The procedure GET-PREDECESSOR-SUCCESSOR-BOTTOM-UP($root(T_i)$) constructs $P_g(x)$ and $S_g(x)$ for each gene x of gene family i in every ancestral genome g, where $root(T_i)$ denotes the root of T_i. Suppose π is the current node and τ and φ are the two direct descendants of π in T_i. Note that either τ or φ might be null.

GET-PREDECESSOR-SUCCESSOR-BOTTOM-UP(π)

```
1    if π is not null and π is non-leaf node
2       then GET-PREDECESSOR-SUCCESSOR-BOTTOM-UP(τ)
3            GET-PREDECESSOR-SUCCESSOR-BOTTOM-UP(φ)
4            h ← φ(π); f ← φ(τ); g ← φ(φ)
5            if ‖ Ãₕ(P_f(τ)) ∩ Ãₕ(P_g(φ)) ‖≠ 0
```

6	**then** $P_h(\pi) \leftarrow \tilde{A}_h(P_f(\tau)) \cap \tilde{A}_h(P_g(\varphi))$
7	**else** $P_h(\pi) \leftarrow \tilde{A}_h(P_f(\tau)) \cup \tilde{A}_h(P_g(\varphi))$
8	**if** $\| \tilde{A}_h(S_f(\tau)) \cap \tilde{A}_h(S_g(\varphi)) \| \neq 0$
9	**then** $S_h(\pi) \leftarrow \tilde{A}_h(S_f(\tau)) \cap \tilde{A}_h(S_g(\varphi))$
10	**else** $S_h(\pi) \leftarrow \tilde{A}_h(S_f(\tau)) \cup \tilde{A}_h(S_g(\varphi))$

The root of the reconciled tree $\mathbb{T}$ is not always the target genome we want to reconstruct. Therefore, we first infer $P_R(x)$ and $S_R(x)$ in the common ancestor R in $\mathbb{T}$. Then we propagate $P_R(i)$ and $S_R(i)$ down the tree until we reach the target ancestor α. We use ADJUST-ANCESTOR-TOP-DOWN to adjust the original $P_g(x_i)$ and $S_g(x_i)$ for every gene x_i in genome g leading from R to α, assuming that the path from R to α has already been recorded (the .next field means the next node on the path from R to α).

ADJUST-ANCESTOR-TOP-DOWN(R, α)
1   $h \leftarrow R;\ g \leftarrow R.next$
2   **while** $h \neq \alpha$
3   **do for** each $x_i \in \mathbf{X}$ where $\mathbf{X} = x_1, -x_1, ..., x_{N_g}, -x_{N_g}$
4   **do if** $\| \tilde{A}_h(P_g(x_i)) \cap P_h(\tilde{A}_h(x_i)) \| \neq 0$
5   **then** $P_g(x_i) \leftarrow P_g(x_i) \cap \tilde{D}_g(\tilde{A}_h(P_g(x_i)) \cap P_h(\tilde{A}_h(x_i)))$
6   **if** $\| \tilde{A}_h(S_g(x_i)) \cap S_h(\tilde{A}_h(x_i)) \| \neq 0$
7   **then** $S_g(x_i) \leftarrow S_g(x_i) \cap \tilde{D}_g(\tilde{A}_h(S_g(x_i)) \cap S_h(\tilde{A}_h(x_i)))$
8           $h \leftarrow g;\ g \leftarrow g.next$

At this point, in the target ancestor α, we have had potential predecessors and successors for each gene. The remaining task is to reconstruct the order based on adjacency information.

2.4   From Ancestral Adjacency to Ancestral Gene Order

We first construct a **predecessor graph** G_α^P and a **successor graph** G_α^S for the target genome α. The digraph $G_\alpha^P = (V, E)$, where $|V| = 2N_\alpha$, is defined such that each gene x_i corresponds to two nodes, i and $-i$, and the set of directed edges is: $E(G_\alpha^P) = \{(u, v) \mid u \in P_\alpha(v)\}$. Similarly, in digraph $G_\alpha^S = (V, E)$, $|V| = 2N_\alpha$, and: $E(G_\alpha^S) = \{(u, v) \mid v \in S_\alpha(u)\}$. Here, (u, v) denotes an arc directed from u to v. Note that an edge in G_α^P is *from* the predecessor, while an edge in G_α^S is *to* the successor. For instance, let g have the chromosome (1 -4 -3 5.1 2). Then G_g^P and G_g^S are as shown in Fig.3(A) and (B), respectively.

We intersect G_α^P and G_α^S, producing the intersection graph $G = G_\alpha^P \cap G_\alpha^S$, retaining edges that are not connecting to either of the endpoints, Φ_A and Φ_Z. Then special care is taken to add endpoint edges, basically retaining all the endpoint edges that appear in both G_α^P and G_α^S. All three graphs (predecessor, successor, and intersection) have the same set of $2N_\alpha$ nodes. G's edges are:

$$E(G) = \left\{ E(G_\alpha^P) \cap E(G_\alpha^S) \right\}$$
$$\cup \left\{ (\Phi_A, v) \mid (\Phi_A, v) \in E(G_\alpha^P) \right\} \cup \left\{ (u, \Phi_Z) \mid (u, \Phi_Z) \in E(G_\alpha^S) \right\} \quad (1)$$

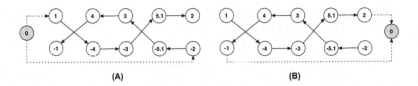

(A)   (B)

Fig. 3. (A) A predecessor graph G_g^P; (B) A successor graph G_g^S

The edges of the intersection graph G indicate consistent predecessor and successor relationships that are supported by $\mathbb{T}$, T_i and the modern genomes. However, they do not necessarily indicate a unique adjacency relationship for a particular gene. Three potential ambiguous cases might occur in the intersection graph, as depicted for node i in Figure 4. In (a), i has several incoming edges. In (b), i has several outgoing edges. In (c), i forms a cycle with j, where each node j satisfies $indegree(j) = outdegree(j) = 1$. (If a more complex cycle exists, then some node falls in either case (a) or case (b)).

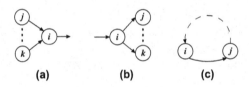

(a)   (b)   (c)

Fig. 4. Three potential ambiguous cases in the intersection graph G

If none of these ambiguous cases is present, the intersection graph itself forms the set of paths that covers all the nodes. In this case, the CARs can be directly defined from this graph as discussed below. When ambiguity exists, we need to resolve the ambiguity and choose appropriate directed edges to form CARs. We assign a weight to each of the directed edges in the remaining graph using the following approach.

For an directed edge (i, j), if $outdegree(i) = 1$ and $indegree(j) = 1$ (in other words, it is not among one of the incoming edges of case (a) nor it is among one of the outgoing edges of case (b)), we set $w_\alpha(i, j) = 1$. Otherwise, the corresponding weight $w_\alpha(i, j)$ is determined recursively by:

$$w_\alpha(i, j) = \frac{d(\alpha, \tau) \cdot w_\varphi(i, j) + d(\alpha, \varphi) \cdot w_\tau(i, j)}{d(\alpha, \tau) + d(\alpha, \varphi)} \tag{2}$$

where $d(\alpha, \tau)$ and $d(\alpha, \varphi)$ are the branch lengths to the left child and right child; $w_\tau(i, j)$ and $w_\varphi(i, j)$ are the edge weights on left child and right child, respectively. On a leaf genome, if (i, j) is present in the predecessor graph, we set $w(i, j) = 1$, otherwise $w(i, j) = 0$. This kind of edge weight can also be determined by a postorder traversal. Note that if an edge (i, j) is involved in ambiguous case (a) or (b), $w(i,j) < 1$. The underlying assumption of equation 2 is that rearrangement is more likely to happen on longer branches.

Our goal is to connect elements into the longest possible CARs that are consistent with the observed data. The problem can be transformed into looking for vertex-disjoint paths that cover all the nodes in the digraph G with the maximum weight. Here we also allow degenerate paths, where there is only one node. The simplified version of this problem when all the edge weights are the same, say 1, is equivalent to the Minimum Path Cover Problem, i.e., finding the minimum number of vertex-disjoint paths covering all the nodes in the digraph. The minimum path cover problem was proved to be NP-hard [18].

We use a greedy approach to achieve an approximate solution, given in the algorithm of FIND-CARS below. We first sort the edges by weight. Then the greedy approach always tries to add the heaviest edge to the resulting path set.

FIND-CARS(G)
1   Sort edges by weight in descending order.
2   Create a new graph C, $V(C) = V(G)$ and $E(C) = \emptyset$
3   **for** each available $(i,j) \in E(G)$, in order of edge weight
4   **do if** $outdegree(i) = 0$ and $indegree(j) = 0$
5   **then** Add edge (i,j) and $(-j,-i)$ to $E(C)$
6   Update $outdegree(i)$ and $indegree(j)$ in C
7   Break cycles in C.
8   **return** C.

Note that the simple greedy process doesn't guarantee there will be no cycle in the path set. We need a final step (line 7) to detect and break the cycles. We use the depth-first-search algorithm to detect cycles in graph G. In fact, we can prove that if there is a cycle, the weight of each edge in that cycle is 1. Therefore, we can simply discard an arbitrary edge to break the cycle (In a variant where circular chromosomes are considered, then cycles would be allowed). The remaining paths in G correspond to the CARs we want to reconstruct.

When adding edges into an existing path, particular care is needed to avoid putting j and $-j$ in the same CAR. In addition, we add both (i,j) and its symmetric version, $(-j,-i)$. For each path found by this approach, a symmetric path in the opposite orientation is also found, since we have nodes for both i and $-i$. The two paths correspond to the same CAR, and eventually we choose one of them.

2.5   Summary

In outline, the whole INFER-CARS-WITH-DUP algorithm can be described as follows, where α is the target ancestor, and $\mathbb{G}$ denotes the collection of modern genomes.

INFER-CARS-WITH-DUP(α)
 1   Construct $\mathbb{T}$ and $\mathbb{T}_x$ (for each gene family x)
 2   $\mathbb{C} \leftarrow$ empty set of CARs
 3   $R \leftarrow root(\mathbb{T})$
 4   Initialize $P_g(i)$ and $S_g(i)$ for each gene i in every g in $\mathbb{G}$
 5   **for** each gene family i
 6   **do** GET-PREDECESSOR-SUCCESSOR-BOTTOM-UP($root(\mathbb{T}_i)$)
 7   ADJUST-ANCESTOR-TOP-DOWN(R, α)
 8   Get graph G according to Equation (1)
 9   $\mathbb{C} \leftarrow$ FIND-CARS(G)
10   **return** $\mathbb{C}$

3   Results

3.1   Simulation Results

We used extensive simulations to test and validate our analysis. The simulator starts with a hypothetical 'ancestor' genome which evolves into the extant species through speciation, inversion, translocation, fusion, fission, insertion, deletion, and duplication. When an operation is applied, the breakpoint is chosen uniformly at random from the set of used or unused breakpoints on this chromosome, depending on the breakpoint reuse ratio. The length of the operation is also picked uniformly at random within the specified distance from the first breakpoint.

We tuned the weights of these operations in order to generate simulated data that makes more biological sense specifically for placental mammalian genomes. The ancestor genome was assigned around 5,000 genes. The parameters or weights of the large scale operations were tuned such that the extant species had around the same number of genes. The breakpoint reuse ratio was kept around 8%-10% and each of the extant species had 5%–10% duplicated genes. We simulated 50 datasets using the phylogenetic tree:
    ((((human,chimp),rhesus),(mouse,rat)),dog).
On average, the ratio of breakpoint reuse is 9.98%, the ratio of duplicated genes in each extant species is 8.12% (rhesus), 7.52% (human), 7.26% (chimp), 7.12% (mouse), 7.85% (rat), and 7.23% (dog), respectively. Also, rearrangements are distributed as 82.33% inversions, 9.40% translocations, 3.86% fusions, and 4.40% fissions. In all the duplication events, 30.40% are tandem duplications and 69.60% are segmental duplications.

We ran our reconstruction program for inferring CARs on each dataset (avg. running time 14.62min) and compared the predicted adjacencies with the known (simulated) ones. Our target ancestor was primate-rodent ancestor and dog was treated as outgroup. For determining the success rate, we considered only the effective ancestral adjacencies (~59% of all ancestral adjacencies) that were broken in at least one lineage in the subtree rooted by primate-rodent ancestor, since the unbroken adjacencies will be found by essentially any procedure.

The frequency of correctly predicted adjacencies was 99.46% (SD=0.43%) for the primate-rodent ancestor. The reconstruction accuracy of human-rhesus ancestor and mouse-rat ancestor is 99.75% (SD=0.27%) and 99.72% (SD=0.25%) respectively.

We did some additional experiments to see how the performance changes in the primate-rodent ancestor if we change parameters in the simulation. We made the effective ancestral adjacency vary by using different number of rearrangement operations. Interestingly, the accuracy didn't change much. For example, when the effective ancestral adjacency is around 10%, the accuracy is 99.67%. When the effective adjacency is around 70%, the accuracy is 99.45%. We think the accuracy didn't really depend on effective adjacency because we used six species in this simulation. We also increased the breakpoint reuse ratio to around 40% when the effective adjacency ratio is 70%, then the accuracy dropped to 96.83%. We concluded from these preliminary experiments that when the number of leaf genomes is reasonable, the reconstruction performance isn't hurt much if we increase the number of operations (as reflected in the effective adjacencies). Instead, the performance will be suffered if we increase the breakpoint reuse ratio to let one ancestral adjacency be broken independently in different lineages.

3.2   Application to Real Data

It has been shown that the unicellular eukaryote *Paramecium tetraurelia*, a ciliate, which contains about 40,000 genes, is a result of at least three whole genome duplication (WGD) with additional rearrangement operations [19]. In that paper, the authors reconstructed the genome architectures of four ancestral genomes, corresponding to the most recent WGD, the intermediary WGD, the old WGD, and the ancient WGD. They used Best Reciprocal Hits to construct a paralogon, which is a pair of paralogous blocks that could be recognized as deriving from a common ancestral region. Then paralogons were merged into single ancestral blocks and the process was iterated until reaching the ancient WGD. However, they didn't intend to figure out the gene orders in each ancestral block. When a paralogon was constructed, the detailed order and orientation of genes inside the block were ignored.

All 39,642 genes form 22,635 gene families (including 11,740 single-gene families), which have been scattered on 676 scaffolds in the present day genome. We tested our algorithm by reconstructing all WGDs except the ancient WGD. We used the gene order in modern *Paramecium tetraurelia* and the gene trees from [19]. The reconciled tree contains one leaf genome, which is the modern genome, as well as ancestral nodes representing duplication events. We built the augmented gene trees accordingly.

Many genes do not have paralogous genes in the paralogons for a particular ancestral genome. If we include all the gene families in the reconstruction, the input data would be very noisy and the resulting CARs would be too fragmented due to the fact that we only have one leaf genome. For example, if we include all the genes, there are 1,937 reconstructed CARs in the old WGD. Therefore, when we were reconstructing CARs in a certain target genome, we did some

Table 1. Number of CARs we reconstructed in three target ancestral genomes

target ancestor	genes we included	anc genes with paralog (from [19])	gene families we used	predicted CARs	anc blocks in [19]
Old WGD	2,981	1,530	559	57	43
Intermediary WGD	11,620	7,996	3,770	144	81
Recent WGD	25,708	24,052	9,951	228	131

preprocessing to only retain genes that have paralogous genes derived from more ancient duplications. Additional genes were also added if their paralogs (from this duplication) were retained in the leaf genome.

For all three genomes, the number of CARs reconstructed by us is greater than the number of ancestral blocks reported in [19] using the paralogon method to construct ancestral blocks. There are two reasons for this: (1) The authors of [19] ignored the gene orders while we take order andorientation into account when inferring CARs. (2) We used more genes in the reconstruction than just the ancestral genes with paralogs, which were essentially used as anchors when building paralogons.

Since paper [19] didn't reconstruct the ancestral gene adjacencies, we couldn't compare our prediction with theirs in detail. Preliminary comparison showed that our prediction has basically and more detailed refinement than the result from [19]. Also, recent studies on genome halving problem [20,21,22,23] might be particularly useful and interesting to be applied to the Paramecium genome. As further ciliate genomes become available, we plan to further investigate the changes of gene orders between different WGDs, using additional outgroup information from closely related species to pick up more adjacencies we couldn't reconstruct now, which will help to determine which methods of reconstructing ancestral architecture are best, and might shed more light on the evolution of the ciliate *Paramecium tetraurelia*.

4   Discussion

In this paper, we extend the method in [8] to reconstruct ancestral gene orders with duplications. We have a simplifying assumption that all the distances in the gene trees are perfect, which makes it easy to reconcile gene trees to the species tree. In reality, we usually have gene trees with approximate distances. Therefore, a more robust reconciliation method is needed, e.g. [24] and [25]. This is a key area for further work.

Our future work will also focus on incorporating the ability to reconstruct evolutionary history with large-scale operations, instead of just figuring out the gene orders. Although solving the median problem is algorithmically challenging, it is completely feasible to provide a plausible history of rearrangements and duplications on each branch in the phylogeny when the descendant genome and the ancestor genome have been both predicted.

Our simulation on large-scale mammalian genome evolution looks promising. However, a number of challenges remain before the genome structure of mammalian ancestors can be accurately predicted in terms of rearrangements and duplications, among which the most difficult would be partitioning the genomes and accurately dating the duplication events.

Acknowledgement

We thank Olivier Jaillon at Centre National de Sequencage in France for providing data of ciliate *Paramecium tetraurelia*.

References

1. Caprara, A.: Formulations and hardness of multiple sorting by reversals. RE-COMB, 84–94 (1999)
2. Pe'er, I., Shamir, R.: The median problems for breakpoints are NP-complete. Electronic Colloquium on Computational Complexity (ECCC), 5(71) (1998)
3. Sankoff, D., Blanchette, M.: Multiple genome rearrangement and breakpoint phylogeny. J. Comput. Biol. 5(3), 555–570 (1998)
4. Moret, B.M.E., Wyman, S.K., Bader, D.A., Warnow, T., Yan, M.: A new implmentation and detailed study of breakpoint analysis. PSB, 583–594 (2001)
5. Bourque, G., Pevzner, P.A.: Genome-scale evolution: reconstructing gene orders in the ancestral species. Genome Res. 12(1), 26–36 (2002)
6. Froenicke, L., Caldes, M.G., Graphodatsky, A., Muller, S., Lyons, L.A., Robinson, T.J., Volleth, M., Yang, F., Wienberg, J.: Are molecular cytogenetics and bioinformatics suggesting diverging models of ancestral mammalian genomes? Genome Res. Genome Res. 16(3), 306–310 (2006)
7. Bourque, G., Tesler, G., Pevzner, P.A.: The convergence of cytogenetics and rearrangement-based models for ancestral genome reconstruction. Genome Res. 16(3), 311–313 (2006)
8. Ma, J., Zhang, L., Suh, B.B., Raney, B.J., Burhans, R.C., Kent, W.J., Blanchette, M., Haussler, D., Miller, W.: Reconstructing contiguous regions of an ancestral genome. Genome Res. 16(12), 1557–1565 (2006)
9. Fitch, W.M.: Toward defining the course of evolution: minimum change for a specific tree topology. Syst. Zool. 20, 406–416 (1971)
10. Rocchi, M., Archidiacono, N., Stanyon, R.: Ancestral genomes reconstruction: An integrated, multi-disciplinary approach is needed. Genome Res. 16(12), 1441–1444 (2006)
11. Eichler, E.E., Sankoff, D.: Structural dynamics of eukaryotic chromosome evolution. Science 301(5634), 793–797 (2003)
12. Sankoff, D.: Genome rearrangement with gene families. Bioinformatics 15(11), 909–917 (1999)
13. Sankoff, D., El-Mabrouk, N.: Duplication, rearrangement and reconciliation. In: Sankoff, D., Nadeau, J.H. (eds.) Comparative genomics: Empirical and analytical approaches to gene order dynamics, map alignment and the evolution of gene families, pp. 537–550. Kluwer Academic Publishers, Dordrecht (2000)

14. Marron, M., Swenson, K.M., Moret, B.M.E.: Genomic distances under deletions and insertions. Theor. Comput. Sci. 325(3), 347–360 (2004)
15. Page, R.D.M., Charleston, M.A.: From gene to organismal phylogeny: reconciled trees and the gene tree/species tree problem. Mol. Phylogenet. Evol. 7(2), 231–240 (1997)
16. Goodman, M., Czelusniak, J., Moore, G.W., Romero-Herrera, A.E., Matsuda, G.: Fitting the gene lineage into its species lineage, a parsimony strategy illustrated by cladograms constructed from Globin Sequences. Syst. Zool. 28(2), 132–163 (1979)
17. Guigo, R., Muchnik, I., Smith, T.F.: Reconstruction of ancient molecular phylogeny. Mol. Phylogenet. Evol. 6(2), 189–213 (1996)
18. Boesch, F.T., Gimpel, J.F.: Covering points of a digraph with point-disjoint paths and its application to code optimization. J. ACM. 24(2), 192–198 (1977)
19. Aury, J.M., Jaillon, O., Duret, L., Noel, B., Jubin, C., Porcel, B.M., Ségurens, B., Daubin, V., Anthouard, V., Aiach, N., et al.: Global trends of whole-genome duplications revealed by the ciliate Paramecium tetraurelia. Nature 444, 171–178 (2006)
20. Seoighe, C., Wolfe, K.H.: Extent of genomic rearrangement after genome duplication in yeast. PNAS 95(8), 4447–4452 (1998)
21. El-Mabrouk, N., Sankoff, D.: The reconstruction of doubled genomes. SIAM J. Comput. 32(3), 754–792 (2003)
22. Alekseyev, M.A., Pevzner, P.A.: Whole genome duplications and contracted breakpoint graphs. SIAM J. Comput. 36(6), 1748–1763 (2007)
23. Zheng, C., Zhu, Q., Sankoff, D.: Genome halving with an outgroup. Evolutionary Bioinformatics 2, 319–326 (2006)
24. Chen, K., Durand, D., Farach-Colton, M.: NOTUNG: a program for dating gene duplications and optimizing gene family trees. J. Comput. Biol. 7(3-4), 429–447 (2000)
25. Bansal, M.S., Burleigh, J.G., Eulenstein, O., Wehe, A.: Heuristics for the gene-duplication problem: A $\Theta(n)$ speed-up for the local search. RECOMB, pp. 238–252 (2007)

Appendix

We discuss in detail the algorithm for determining the reconciled tree and augmented gene tree. Let S be a rooted species tree and A be an unrooted gene tree. We assume that S has an infinitely long incoming edge leading into its root to accommodate ancient duplications, if needed. A **reconciliation** of A with S is a mapping ϕ from the nodes of A into the set of nodes and points along the edges of S with the following properties: (1) Every leaf l of A maps to a leaf $\phi(l)$ of S of the same species; (2) Each internal node a of A maps to a point $\phi(a)$ in S that lies either at a node or at a point on an edge in S; and (3) The mapping ϕ is isometric in the sense that for every leaf node l in A, the distance from a to l in A is the same as the distance from $\phi(a)$ to $\phi(l)$ in S. When $\phi(a)$ is a node in the species tree S, we say that a is a *speciation node* in A, and when $\phi(a)$ is a point that lies along an edge in S we say that a is a *duplication node* in A and we create a corresponding duplication node at $\phi(a)$ in S.

Any internal node x in the unrooted binary tree A will be connected to three other nodes u, v, and w, defining three possible rooted and directed subtrees U,

V, and W of A, respectively. If A is to be successfully reconciled with S, two of these subtrees, say U and V, must map to directed subtrees of S in such a way that $\phi(x)$ lies above $\phi(u)$ and $\phi(v)$. To define the complete reconciliation, we proceed inductively, assuming that we have already reconciled subtrees U and V of A, and extending this reconciliation to include x. Let d_1' and d_2' be the distances in A from x to u and v to x, respectively. Let $\tilde{x}$ be the last common ancestor of $\phi(u)$ and $\phi(v)$ in S. Let d_1 and d_2 be the distances in S from $\phi(u)$ and $\phi(v)$ to $\tilde{x}$, respectively. We will have $d = d_1' + d_2' - d_1 - d_2 \geq 0$. Then the subtree of A rooted at x and containing U and V can be reconciled with S by extending the reconciliation of its subtrees U and V by adding a point $\phi(x)$. The point $\phi(x)$ must lie at a distance $d/2$ upstream from $\tilde{x}$ in S, along the unique path in S leading into $\tilde{x}$. Such a point always exists in S because we have added an infinitely long stem branch leading into the original root of S. If $d = 0$, then $\phi(x) = \tilde{x}$. It is clear that the distance from $\phi(x)$ to $\phi(l)$ for any leaf l in U or V must be correct, since the distances from $\phi(u)$ and $\phi(v)$ are correct by the inductive hypothesis, and the additional distance added from $\phi(u)$ or $\phi(v)$ to $\phi(x)$ in the above construction is exactly the increment needed to keep the distances correct.

So long as A has more than one node, the inductive construction terminates with two adjacent nodes y and z that dominate all other nodes in A, in the sense that both the subtree Y rooted at y and pointing away from z, and the subtree Z rooted at z and pointing away from y, are reconciled into subtrees of S. Then the final step is to determine the root of the gene tree. Now let d be the distance between y and z in A. Let $\tilde{r}$ be the last common ancestor of $\phi(y)$ and $\phi(z)$ in S. Let d_1 and d_2 be the distances in S from $\phi(y)$ and $\phi(z)$ to $\tilde{r}$, respectively. We define the root r of the gene tree A as the point at distance $(d + d_1 - d_2)/2$ from y and $(d + d_2 - d_1)/2$ from z along the edge connecting y and z, and $\phi(r)$ as the corresponding point at distance $(d - d_1 - d_2)/2$ upstream from $\tilde{r}$ in S. This completes the reconciliation.

We construct the reconciled tree by repeating the above procedure for each gene tree. Each reconciliation adds new duplication nodes to S until the final reconciled tree $\mathbb{T}$ is built.

Reconstructing an Inversion History in the *Anopheles Gambiae* Complex

Ai Xia, Maria V. Sharakhova, and Igor V. Sharakhov

Department of Entomology, Virginia Tech, Blacksburg, VA 24061, USA
igor@vt.edu

Abstract. The phylogenetic relationships among the members of species complexes can be inferred from the distribution of fixed inversions if outgroup arrangements are known. The *Anopheles gambiae* complex consists of seven African mosquito species that can be differentiated based on ten fixed inversions. However, the phylogenetic relationships among the members remain unclear. This paper demonstrates that physical maps of the outgroup species *A. funestus* and *A. stephensi* can be used for determining ancestral chromosome arrangements in the *A. gambiae* complex. Gene order comparisons have been performed using the Multiple Genome Rearrangements (MGR) and Sorting Permutation by Reversals and block-INterchanGes (SPRING) programs. The analysis has identified the chromosomal arrangements which are likely to be the ancestral in the complex.

Keywords: *Anopheles gambiae* complex, chromosome rearrangement, phylogeny.

1 Introduction

Anopheles gambiae, the most important malaria vector in the world, belongs to a complex of seven sibling species, the group of closely related species, which cannot be distinguished morphologically but can be differentiated based on chromosomal arrangements and molecular markers. Most species of the *A. gambiae* complex are of less or no importance as malaria vectors. Identification of the member that is the closest to the ancestral species for the *A. gambiae* complex will provide a framework for determining the evolutionary genomic changes associated with the increased ability to transmit a malaria parasite.

The karyotype of malaria mosquitoes consists of 3 pairs of chromosomes: one pair of acrocentric sex chromosomes X (X and Y in males), and two pairs of submetacentric autosomes 2 and 3. There are ten fixed inversions in the *A. gambiae* complex. *A. quadriannulatus* A and *A. quadriannulatus* B have the standard (presumably ancestral) chromosomal arrangements [1]. The notation for the standard karyotype is X+, 2R+, 2L+, 3R+, 3L+. The other members of the complex have fixed inversions on various chromosomal arms. The 2La inversion (the inverted arrangement on the left arm of the chromosome 2) is fixed in *A. arabiensis* and *A. merus*, but is polymorphic in *A. gambiae* [1, 2]. *A. merus* and *A. gambiae* share the Xag inversion, while *A. arabiensis* has the Xbcd inversion. Additionally, *A. merus*

G. Tesler and D. Durand (Eds.): RECOMB-CG 2007, LNBI 4751, pp. 136–148, 2007.

and *A. gambiae* differ from each other by two overlapping inversions on 2R, "o" and "p". *A. bwambae* and *A. melas* share the 3La arrangement while *A. melas* carries a 2Rm inversion [1].

The reconstruction of an inversion history is based on two principles: monophyly and parsimony. For a long time, the standard chromosomal arrangements of *A. quadriannulatus* (homosequential species A and B) had been considered the closest to the ancestral species because of their central position relative to other species in the complex [1, 3] (Fig. 1A). However later *A. arabiensis*, was assumed to be the closest to the ancestral species, in part because it has a fixed 2La inversion cytologically identified in two members of the *Anopheles subpictus* complex, i.e., outside of the *A. gambiae* complex [4] (Fig. 1B). Recently, the analysis of the inversion breakpoint structure revealed that *A. arabiensis*, *A. gambiae*, and *A. merus* share the same 2La arrangement. Moreover, the molecular features of the breakpoints strongly suggested that this arrangement is ancestral [2]. It has been proposed that a possible ancestor of the complex, *A. arabiensis*, may have originated in the Middle East and reached Africa through the arid Arabian peninsula [4]. However, because 2La is present in *A. arabiensis*, *A. gambiae*, and *A. merus*, any of these three species can now be considered the closest to the ancestral species (Fig. 1).

Additional studies are required in order to reconstruct the phylogenetic relationships among the members of the *A. gambiae* complex. Specifically, the relationships of *A. quadriannulatus*, *A. arabiensis*, *A. gambiae*, and *A. merus* among each other and with the ancestral species to the complex remain unresolved. One way to resolve these phylogenetic relationships is to determine ancestry of inversion arrangements on the chromosomes X, 2L, and 2R. A reconstruction of the *A. gambiae* complex phylogeny using polytene chromosome maps of outgroup species has been attempted [5]. Although, the sister group relationships *A. bwambae* + *A. melas* and *A. gambiae* + *A. merus* have been confirmed, the identification of the ancestral arrangements have failed because a cytogenetic map provides insufficient information on gene order.

This paper determines ancestry of the gene order arrangements of two chromosomal arms in the *A. gambiae* complex using physical maps of the outgroup species *A. funestus* [6, 7] and *A. stephensi* (developed in this study). Thirty six probes have been used for development of physical maps for the *A. stephensi* 2R and 3L chromosomes corresponding to the 2R and 2L arms of *A. gambiae* and to the 2R and 3R arms of *A. funestus*. The three species belong to different series within the subgenus Cellia: Pyretophorus (*A. gambiae*), Myzomyia (*A. funestus*), and Neocellia (*A. stephensi*) [8]. The *A. gambiae* and *A. funestus* lineages diverged from a common ancestor at least 36 million years ago [9]. *A. gambiae* is more closely related to *A. stephensi* than to *A. funestus* [10, 11]. In this study, we have used the Multiple Genome Rearrangements (MGR) [12] and Sorting Permutation by Reversals and block-INterchanGes (SPRING) [13] programs to calculate inversion distances among the species. We have confirmed the ancestral state of the 2La arrangement and have shown that the 2Rop arrangement is more likely to be ancestral in the complex than the 2R+ arrangement.

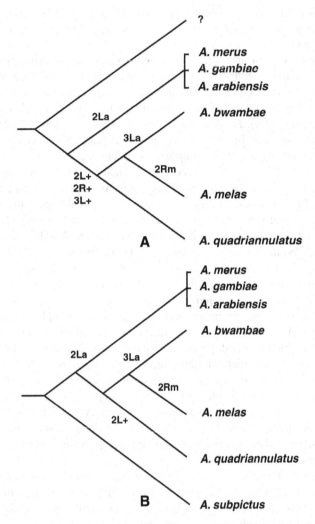

Fig. 1. The hypothetical phylogenetic trees of the *A. gambiae* complex based on the central position of the *A. quadriannulatus* chromosomal arrangements in the complex (A) and based on the ancestry of the 2La arrangement (B). The known chromosomal arrangements that support these trees are shown. The question mark indicates that an outgroup species that supports this three is not known.

2 Materials and Methods

Mosquito Strain and Chromosome Preparation. The Indian Wild Type strain of *A. stephensi*, which is a standard laboratory strain [14, 15], was used in this study. Ovaries were extracted from half-gravid females and preserved in Carnoy's fixative solution (3 ethanol: 1 glacial acetic acid by volume). Then they were fixed for 24 hours at room temperature and dissected in 50% propionic acid. A cover slide was placed on the follicles and pressed to extrude the cells. The banding pattern of polytene chromosomes was examined using an Olympus phase-contrast microscope (x10 objective). Slides with good chromosomal preparations were dipped in liquid nitrogen, then the cover slips were removed and slides were dehydrated in 50%, 70%, 90%, and 100% ethanol.

Probe Preparation and Fluorescent *In Situ* Hybridization. Thirty six conserved *A. funestus* and *A. gambiae* cDNA clones, as well as *A. gambiae* BAC clones (Table 1), were mapped to *A. stephensi* polytene chromosomes 2R and 3L using FISH (Fluorescent *In Situ* Hybridization). Development of physical maps for other chromosomes is in progress. The *A. gambiae* cDNAs of A.Gam.ad.cDNA1 and A.Gam.ad.cDNA.blood1 libraries [16] as well as the *A. gambiae* BAC clones of NotreDame1 [16] and ND-TAM [17] libraries were obtained from the Malaria Research and Reference Reagent Resource Center (MR4) (www.mr4.org). The *A. funestus* cDNAs derived from the *A. funestus* SMART Library [6]. Recombinant cDNA and BAC clones were isolated using a Sigma Kit (Sigma). Genomic inserts from the SMART cDNA library were PCR amplified using T3/T7 or Amplimer primers in following conditions: 95°C for 5 min; 25 cycles of 94°C for 30 s, 70°C for 2 min,68°C for 3 min with the Amplimer primer and 95°C for 5 min; and 25 cycles of 94°C for 30 s, 50°C for 30 min, 72°C for 30s, 72°C for 5 min with the T3/T7 primers. The DNA was labeled with Cy5-AP3-dUTP (GE Healthcare UK Ltd, Buckinghamshire, England) by Random Primers DNA Labeling System (Invitrogen Corporation, Carlsbad, CA, USA) or the nick translation kit (Amersham, Bioscience, Little Chalfont Buckinghamshire).

DNA probes were hybridized to the chromosomes at 39°C overnight in hybridization solution (Invitrogen Corporation, Carlsbad, CA, USA). Then, the chromosomes were washed in 0.2XSSC (Saline-Sodium Citrate: 0.03M Sodium Chloride, 0.003M Sodium Citrate), counterstained with YOYO-1, and mounted in DABCO. Fluorescent signals were detected and recorded using a Zeiss LSM 510 Laser Scanning Microscope (Carl Zeiss MicroImaging, Inc., Thornwood, NY, USA). Localization of a signal was accomplished using a standard cytogenetic map for *A. stephensi* [18]. No variation in signal localization was detected among all the nuclei examined for a given probe.

Computational Methods and Analysis. Analysis included 36 probes uniquely located on the *A. stephensi* and *A. funestus* chromosomes [6, 7]. Locations of the *A. funestus* and *A. stephensi* sequences in the *A. gambiae* genome were determined by BLASTN with default parameters using the VectorBase website (http://www.vectorbase.org/Tools/BLAST/#). Because these lineages diverged at least 36 million years ago [9] we considered only hits with e-values less than e^{-20} and alignments longer than 100 nt for ESTs and with e-values less than e^{-5} and alignments longer than 40 nt for microsatellite containing noncoding sequences. Chromosomal locations of the *A. gambiae* BAC clones were found using the search option on the VectorBase website (http://www.vectorbase.org/Search/Keyword/).

The calculation of inversion distances among species of the *A. gambiae* complex, as well as those for *A. funestus* and *A. stephensi* were performed using MGR and SPRING programs. The Multiple Genome Rearrangements (MGR) program is available at www.cs.ucsd.edu/groups/bioinformatics/MGR. This program implements an algorithm which seeks a tree that minimizes the sum of the rearrangements over all the edges of the tree [12]. The Sorting Permutation by Reversals and block-INterchanGes (SPRING) program is available at http:// algorithm.cs.nthu.edu.tw/ tools/SPRING/index.php. SPRING computes both the breakpoint and rearrangement distances between any pair of two chromosomes [13]. It also shows phylogenetic trees that are reconstructed based on the rearrangement and breakpoint distance matrixes. The algorithms of MRG and SPRING are different. MRG uses heuristic strategies to reconstruct a phylogenetic tree of input species. SPRING uses the Neighbor-Joining method to reconstruct a tree instead of a heuristic method. We have chosen these two methods for our analysis because they use gene order information as opposed to nucleotide sequences.

3 Results

Inversion Distances in the *A. gambiae* Complex. There are 10 fixed inversions in the *A. gambiae* complex [1]. We used information on positions of the inversion breakpoints available from Coluzzi et al. [1] to identify an order of 30 cytological markers in all seven species. These markers are the patterns of chromosomal bands and interbands associated with each breakpoint. We were able to determine the sign of these markers based on the banding pattern. The two outgroups are not present here because not all the *A. stephensi* and *A. funestus* chromosomal arms have physical maps of sufficient density. The following gene orders were entered into the MGR and SPRING programs:

>*A. gambiae* 2R+ 2La 3L+ Xag
1 2 3 4 5 6 7 8 9 10 11 12 $
13 14 $
15 16 $
-28 -27 -26 -25 -24 -23 -22 -21 -20 -19 17 18 29 30 $
>*A. arabiensis* 2R+ 2La 3L+ Xbcd
1 2 3 4 5 6 7 8 9 10 11 12 $
13 14 $
15 16 $
17 18 19 20 -26 -25 -30 -29 -28 -27 21 22 -24 -23 $
>*A. merus* 2Rop 2La 3L+ Xag
1 -7 -6 -5 -4 -9 -8 2 3 10 11 12 $
13 14 $
15 16 $
-28 -27 -26 -25 -24 -23 -22 -21 -20 -19 17 18 29 30 $
>*A. melas* 2Rm 2L+ 3La X+
1 2 3 4 5 -11 -10 -9 -8 -7 -6 12 $
14 -13 $
16 -15 $
17 18 19 20 21 22 23 24 25 26 27 28 29 30 $
>*A. quadriannulatus* (species A and B) 2R+ 2L+ 3L+ X+

1 2 3 4 5 6 7 8 9 10 11 12 $
14 -13 $
15 16 $
17 18 19 20 21 22 23 24 25 26 27 28 29 30 $
>*A. bwambae* 2R+ 2L+ 3La X+
1 2 3 4 5 6 7 8 9 10 11 12 $
14 -13 $
16 -15 $
17 18 19 20 21 22 23 24 25 26 27 28 29 30 $

Both MGR and SPRING programs recovered all 10 inversions and identified inversion distances among all species correctly. An unrooted tree produced by MGR can serve as a working hypothesis for determining phylogenetic relations in the complex (Fig. 2).

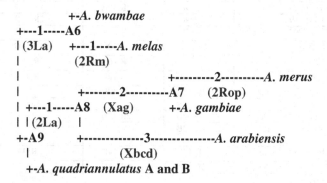

Fig. 2. An unrooted tree of the *A. gambiae* complex recovered by MGR program. The number of rearrangements that occurred on each edge is shown. The names of fixed inversions are shown in parentheses.

Physical Maps for the *A. stephensi* Chromosomes. Table 1 shows a list of DNA probes used for the physical mapping of 2R and 3L chromosomes of *A. stephensi*. Also, it includes the probes mapped to *A. funestus* chromosomes previously [6, 7]. These data reveal the feasibility of interspecific *in situ* hybridization with the *A. stephensi* chromosomes and support the cytological observations about the whole arm translocations in subgenus *Cellia* [8]. Accordingly, chromosomes 2R are homologous across all three species. The 2L arm of *A. gambiae* corresponds to the 3R of *A. funestus* and the 3L of *A. stephensi*.

The same cDNA markers from the *A. funestus* SMART library were hybridized to chromosomes of both species. In some cases, *A. gambiae* cDNA and BAC clones, that contain sequences homologous to *A. funestus* cDNAs, were mapped to *A. stephensi* chromosomes. Resolution of the physical maps for *A. stephensi* chromosomes was 2.8 Mb for 2R and 3.5 Mb for 3L.

Table 1. Comparative cytological positions of DNA probes mapped to chromosomes in *A. gambiae*, *A. funestus*, and *A. stephensi*. The asterisk (*) indicates the major BLASTN hit in *A. gambiae*.

				Chromosomal location		
	Probes on 2R and 2L of *A. gambiae*					
	Probe	Accession	e-value	*A. gambiae*	*A. funestus*	*A. stephensi*
1	21_F03	BU038956	1e-23	2R:7A	2R:7B	
	04L11	AL141975		2R:7A		2R:7A
2	01_H04	BU038873	1e-165	2R:7B	2R:7A	2R:7B
3	21_F12	BU038958	2e-69	2R:8A	2R:12D	2R:13A
4	36_B06	BU038996	3e-80	2R:8D	2R:8E	
	105H10	BH368219		2R:8D		2R:9C
5	12_G10	BU038913	1e-105	2R:8E	2R:15C	2R:8A
6	AFND5	AF171035	3e-18	2R:9B	2R:15B	
	25P09	AL153306		2R:9B		2R:10D
7	FUN O	AY116019	3e-05	2R:9C	2R:18A	
	11A13	AL145719		2R:9C		2R:14B
8	11_D03	BU038903	1e-60	2R:10A	2R:9A	2R:10A
9	04_D06	BU038877	6e-35	2R:11A	2R:10C	2R:16AB
10	08_E06	BU038895	4e-61	2R:11C	2R:16A	2R:10D
11	25_E09	BU038972	3e-83	2R:12B	2R:12B	2R:18B
12	13_F11	BU038919	6e-57	2R:12B	2R:12B	
	129M18	BH377340		2R:12B		2R:18B
13	15_F10	BU038925	9e-65	2R:12B	2R:12B	
	196004496 26240	BM655548	1e-136	2R:12B		2R:18B
14	06_B01	BU038882	1e-90	2R:12D	2R:14D	2R:9A
15	03_D09	BU038874	5e-27	2R:13E	2R:17C	
	31M01	AL611707		2R:13E		2R:8C
16	12_G11	BU038914	3e-40	2R:15B	2R:18C	2R:14C
17	12_H09	BU038915	5e-82	2R:15D	2R:18D	2R:11C
18	29_F03	BU038988	5e-48	2R:15D	2R:11B	
	169F11	BH369697		2R:15D		2R:19A
19	11_E07	BU038905	1e-131	2R:16A	2R:14C	2R:19A
20	13_C03	BU038918	3e-55	2R:17C	2R:13C	
	08O05	AL144514				2R:17A
21	18_D12	BU038940	7e-71	2R:18C	2R:14B	2R:12C
22	11_B04	BU038900	1e-132	2R:19C	2R:19C	2R:19BC
23	28_C07	BU038985	4e-67	2L:20C	3R:35C	
	101C3	BH388218		2L:20C		3L:38B
24	04_D07	BU038878	8e-30	2L:21A	3R:35A	
	02A19	AL140406		2L:21A		3L:39D
25	95_H01	BU039015	4e-27	2L:22D	3R:31D	
	27O10	AL154432				3L:42C
26	AFND19	AF171049	2e-62	2L:22F	3R:34A	
	131F22	BH390198				3L:44C
27	11_F09	BU038906	1e-100	2L:26B	3R:35F	3L:45C
28	09_C11	BU038897	1e-63	2L:26A	3R:36C	3L:43C
29	16_F07	BU038931	1e-129 4e-57	2L:25D-26A* 2L:25A	3R:36E	3L:42A
30	21_E03	BU038955	2e-35	2L:24C	3R:35F	

Table 1. *(continued)*

	140N16	BH384642		2L:24C		3L:39C
31	61_E02	BU039003	9e-86	2L:23D	3R:30C	3L:45A
32	36_A10	BU038993	8e-25	2L:23C	3R:35F	
	150F12	BH385494		2L:23C		3L:44A
33	66_E11	BU038987	1e-134	2L:23A	3R:33D	
	196004497 16320	BM606621		2L:23A		3L:40A
34	08_B09	BU038894	1e-48	2L:26D	3R:30C	
	04C08	AL607764		2L:26D		3L:45A
35	06_G08	BU038889	5e-56	2L:27A	3R:30C	3L:46D
36	18_G01	BU038941	7e-69	2L:28C	3R:29B	3L:46A

The Ancestral State of the 2La Arrangement. Traditionally, the 2La inversion was considered to be derived from the 2L+ standard arrangement [1, 3]. This view was questioned because of the presence of 2La arrangement in the Oriental *A. subpictus* complex [4] and full-length genes and their pseudogene copies only at breakpoints of the 2L+ arrangement [2]. To assume this breakpoint structure ancestral would require the improbable generation of two pseudogene copies before rearrangement at distant reciprocal locations on the 2L+ chromosome, followed by their exact excision during generation of the inversion.

To confirm the ancestral gene order in the 2L arm, inversion distances were calculated among the *A. merus* 2La arm, the *A. quadriannulatus* 2L+ arm, the *A. funestus* 3R arm, and the *A. stephensi* 3L arm using 14 uniquely located probes. The physical maps do not provide the signs of the DNA markers on the *A. funestus* and *A. stephensi* chromosomes. The gene orders for these arms were as follows:

> *A. quadriannulatus* 2L+
1 2 3 4 5 6 7 8 9 10 11 12 13 14 $
> *A. merus* 2La
1 2 3 4 -11 -10 -9 -8 -7 -6 -5 12 13 14 $
> *A. funestus* 3R
7 6 8 5 10 1 2 4 11 3 9 12 13 14 $
> *A. stephensi* 3L
1 2 8 11 3 7 6 4 10 5 9 12 13 14 $

Figure 4 shows that the gene order of 2La, and not that of 2L+, was closer to gene orders of *A. funestus* 3R and *A. stephensi* 3L. This result provides independent evidence of the ancestral state of the 2La arrangement in *A. gambiae*. SPRING produced the same tree based on rearrangements.

More conclusive evidence occurs based on the *in situ* hybridization of DNA probes from the 2L+a proximal breakpoint of *A. gambiae* to the chromosomes of *A. stephensi*. The 146D17 BAC clone spanning the 2L+ proximal breakpoint hybridized to two locations, 40A and 44C, on the chromosome 3L of *A. stephensi*. The 131F22 BAC clone partly overlaps with 146D17, hybridized in only a single location, 44C (Fig. 3).

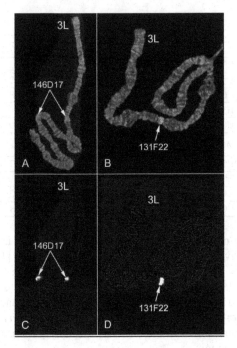

Fig. 3. Fluorescent *in situ* hybridization of 146D17 labeled with Cy5 (A, C) and 131F22 labeled with Cy3 (B, D) performed on the chromosomes of *A. stephensi*. Arrows point at the hybridization signals. A and B show the banding pattern of the chromosomes counterstained with the fluorophore YOYO-1. C and D show fluorescence due to hybridization

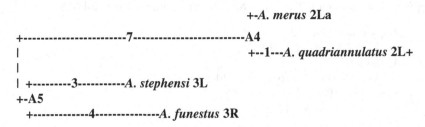

Fig. 4. A tree recovered by MGR program. The number of rearrangements that occured on each edge is shown.

Follow-up experiments involving BAC fragments derived from both sides of the breakpoint yielded only single sites of hybridization for each fragment, either 40A or 44C (data not shown). These results indicate that the breakpoint structure of the 2L+ arrangement is not present in the outgroup species *A. stephensi* and, therefore, is not likely to be ancestral. The derived nature of the 2L+ arrangement and the ancestral status of 2La suggest that *A. arabiensis*, *A. gambiae*, or *A. merus* must be considered the closest to the ancestral species.

The Possible Ancestral State of the 2Rop Arrangement. *A. merus* and *A. gambiae* share the Xag inversion, *A. arabiensis* has the Xbcd inversion, and the other members of the complex have the "standard" X arrangement. Unfortunately, the physical maps of *A. funestus* and *A. stephensi* provided insufficient markers to cover all five inversions. Therefore, the ancestral state of X chromosome cannot be conclusively determined.

A. merus and *A. gambiae* differ from each other by two overlapping inversions on 2R: "o" and "p". We used 22 uniquely located markers that were common for the *A. funestus* and *A. stephensi* maps. The physical maps do not provide the signs of the DNA markers on the *A. funestus* and *A. stephensi* chromosomes. The gene orders for these arms were as follows:

>*A. gambiae* 2R+
1 2 3 4 5 6 7 8 9 10 11 12 13 14 15 16 17 18 19 20 21 22 $
>*A. merus* 2Rop
1 2 3 4 5 -14 -13 -12 -11 -10 -9 -8 -15 6 7 16 17 18 19 20 21 22 $
>*A. funestus* 2R
2 1 4 8 9 18 11 12 13 3 20 21 19 14 6 5 10 15 7 16 17 22 $
>*A. stephensi* 2R
1 2 5 15 14 4 8 10 6 17 21 3 7 16 9 20 11 12 13 18 19 22 $

The MGR and SPRING programs identified both inversions that separate *A. merus* and *A. gambiae* and determined that the 2Rop typical for *A. merus* arrangement is ancestral (Fig. 5).

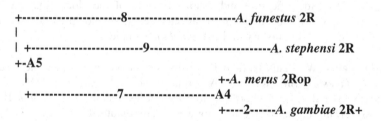

Fig. 5. An MGR phylogenetic tree based on gene orders on 2R arms

4 Conclusion

The analysis reported in this paper showed that physical maps of outgroup species can be used for reconstructing phylogenies based on gene orders within species complexes. No contradictions were found between *A. gambiae* complex phylogenies when the *A. stephensi* or *A. funestus* physical maps were used as outgroups. Our analysis confirmed the ancestral status of 2La postulated in previous studies [2, 4]. Moreover, the *in situ* hybridization of DNA probes from the 2L+a proximal breakpoint of *A. gambiae* to the chromosomes of *A. stephensi* has provided conclusive evidence for the ancestry of the 2La arrangement. The derived nature of the 2L+ inversion leads to a revised history of the *A. gambiae* complex in which *A. arabiensis*,

A. gambiae, A. merus, and not *A. quadriannulatus* must be considered the closest to the ancestral species (Fig. 1B).

The relatively high density of uniquely located markers on the 2R physical map determined the ancestral status of 2Rop arrangement of *A. merus*. These results suggest that the breakpoint structure of the 2Rop and not 2R+ arrangement is present in outgroup species. It is possible to test this hypothesis by mapping DNA probes from the 2Rop breakpoints of *A. merus* to the chromosomes of *A. stephensi*. If the ancestral status of 2Rop is confirmed at the molecular level, *A. merus* must be considered the closest to the ancestral species. This East African saltwater mosquito is not a principal vector of human malaria. However, its important role in malaria transmission in Madagascar has been documented [19]. *A. merus* differs from *A. gambiae*, the most important malaria vector in the world, by only two fixed inversions on 2R. Of eight polymorphic inversions described in *A. gambiae s.s.*, seven occur on chromosome 2R and one on 2L [1]. These inversions are associated with epidemiologically important ecological adaptations, such as tolerance to aridity [1, 3]. Generation of many of these important inversions would have been impossible without initial fixation of the 2R+ arrangement.

The X chromosome had the highest rate of inversion fixation in the complex. The monophyletic origin of the Xag arrangement, common to *A. merus* and *A. gambiae*, has been supported by the molecular analysis of DNA sequences from regions inside Xag [20, 21]. *A. arabiensis* has the Xbcd arrangement. Marker densities on the X chromosome physical maps for *A. stephensi* or *A. funestus* were insufficient to detect all five inversions. Development of a higher resolution physical map for the *A. stephensi* X chromosome will lead to the determination of the ancestral status among X+, Xag, and Xbcd. Identification and characterization of the clones that contain breakpoints for Xag, Xbcd, and 2Rop arrangements should provide detailed information on the inversion history in the *A. gambiae* complex.

Acknowledgements. We thank Nora J. Besansky for providing the genomic inserts from the *A. funestus* SMART cDNA library. The *A. gambiae* cDNAs of A.Gam.ad.cDNA1 and A.Gam.ad.cDNA.blood1 libraries and the *A. gambiae* BAC clones of NotreDame1 and ND-TAM libraries were obtained from the Malaria Research and Reference Reagent Resource Center (MR4). We thank the reviewers for suggestions to improve the manuscript and Ying Chig Lin for useful discussion on MGR and SPRING programs. This study was supported by Agricultural Experimental Station and Fralin Biotechnology Center at Virginia Tech.

References

1. Coluzzi, M., Sabatini, A., della Torre, A., Di Deco, M.A., Petrarca, V.: A polytene chromosome analysis of the Anopheles gambiae species complex. Science 298, 1415–1418 (2002)
2. Sharakhov, I.V., White, B.J., Sharakhova, M.V., Kayondo, J., Lobo, N.F., Santolamazza, F., Della Torre, A., Simard, F., Collins, F.H., Besansky, N.J.: Breakpoint structure reveals the unique origin of an interspecific chromosomal inversion (2La) in the Anopheles gambiae complex. Proc. Natl. Acad. Sci. U S A 103, 6258–6262 (2006)

3. Coluzzi, M., Sabatini, A., Petrarca, V., Di Deco, M.A.: Chromosomal differentiation and adaptation to human environments in the Anopheles gambiae complex. Trans. R Soc. Trop. Med. Hyg. 73, 483–497 (1979)

4. Ayala, F.J., Coluzzi, M.: Chromosome speciation: humans, Drosophila, and mosquitoes. Proc. Natl. Acad. Sci. USA 1, 6535–6542 (2005)

5. Pape, T.: Cladistic analyses of mosquito chromosome data in Anopheles subgenus Cellia (Diptera: Culicidae). In: Mosq. Syst., pp. 241–211 (1992)

6. Sharakhov, I.V., Serazin, A.C., Grushko, O.G., Dana, A., Lobo, N., Hillenmeyer, M.E., Westerman, R., Romero-Severson, J., Costantini, C., Sagnon, N., Collins, F.H., Besansky, N.J.: Inversions and gene order shuffling in Anopheles gambiae and A funestus. Science 298, 182–185 (2002)

7. Sharakhov, I., Braginets, O., Grushko, O., Cohuet, A., Guelbeogo, W.M., Boccolini, D., Weill, M., Costantini, C., Sagnon, N., Fontenille, D., Yan, G., Besansky, N.J.: A microsatellite map of the African human malaria vector Anopheles funestus. J. Hered 95, 29–34 (2004)

8. Green, C., Hunt, R.: Interpretation of variation in ovarian polytene chromosomes of Anopheles funestus Giles, A. parensis Gillies, and A. aruni? Genetica 51, 187–195 (1980)

9. Krzywinski, J., Grushko, O.G., Besansky, N.J.: Analysis of the complete mitochondrial DNA from Anopheles funestus: an improved dipteran mitochondrial genome annotation and a temporal dimension of mosquito evolution. Mol. Phylogenet Evol. 39, 417–423 (2006)

10. Harbach, R.E., Kitching, I.J.: Reconsideration of anopheline mosquito phylogeny (Diptera: Culicidae: Anophelinae) based on morphological data. Systematics and Biodiversity 3, 345–374 (2005)

11. Marshall, J.C., Powell, J.R., Caccone, A.: Short report: Phylogenetic relationships of the anthropophilic Plasmodium falciparum malaria vectors in Africa. Am. J. Trop. Med. Hyg. 73, 749–752 (2005)

12. Bourque, G., Pevzner, P.A.: Genome-scale evolution: reconstructing gene orders in the ancestral species. Genome Res. 12, 26–36 (2002)

13. Lin, Y.C., Lu, C.L., Liu, Y.C., Tang, C.Y.: SPRING: a tool for the analysis of genome rearrangement using reversals and block-interchanges. Nucleic Acids Res. 34, W696–699 (2006)

14. Lim, J., Gowda, D.C., Krishnegowda, G., Luckhart, S.: Induction of nitric oxide synthase in Anopheles stephensi by Plasmodium falciparum: mechanism of signaling and the role of parasite glycosylphosphatidylinositols. Infect. Immun. 73, 2778–2789 (2005)

15. Luckhart, S., Rosenberg, R.: Gene structure and polymorphism of an invertebrate nitric oxide synthase gene. Gene 232, 25–34 (1999)

16. Holt, R.A., Subramanian, G.M., Halpern, A., Sutton, G.G., Charlab, R., Nusskern, D.R., Wincker, P., Clark, A.G., Ribeiro, J.M., Wides, R., Salzberg, S.L., Loftus, B., Yandell, M., Majoros, W.H., Rusch, D.B., Lai, Z., Kraft, C.L., Abril, J.F., Anthouard, V., Arensburger, P., Atkinson, P.W., Baden, H., de Berardinis, V., Baldwin, D., Benes, V., Biedler, J., Blass, C., Bolanos, R., Boscus, D., Barnstead, M., Cai, S., Center, A., Chaturverdi, K., Christophides, G.K., Chrystal, M.A., Clamp, M., Cravchik, A., Curwen, V., Dana, A., Delcher, A., Dew, I., Evans, C.A., Flanigan, M., Grundschober-Freimoser, A., Friedli, L., Gu, Z., Guan, P., Guigo, R., Hillenmeyer, M.E., Hladun, S.L., Hogan, J.R., Hong, Y.S., Hoover, J., Jaillon, O., Ke, Z., Kodira, C., Kokoza, E., Koutsos, A., Letunic, I., Levitsky, A., Liang, Y., Lin, J.J., Lobo, N.F., Lopez, J.R., Malek, J.A., McIntosh, T.C., Meister, S., Miller, J., Mobarry, C., Mongin, E., Murphy, S.D., O'Brochta, D.A., Pfannkoch, C., Qi, R., Regier, M.A., Remington, K., Shao, H., Sharakhova, M.V., Sitter, C.D., Shetty, J.,

Smith, T.J., Strong, R., Sun, J., Thomasova, D., Ton, L.Q., Topalis, P., Tu, Z., Unger, M.F., Walenz, B., Wang, A., Wang, J., Wang, M., Wang, X., Woodford, K.J., Wortman, J.R., Wu, M., Yao, A., Zdobnov, E.M., Zhang, H., Zhao, Q., Zhao, S., Zhu, S.C., Zhimulev, I., Coluzzi, M., della Torre, A., Roth, C.W., Louis, C., Kalush, F., Mural, R.J., Myers, E.W., Adams, M.D., Smith, H.O., Broder, S., Gardner, M.J., Fraser, C.M., Birney, E., Bork, P., Brey, P.T., Venter, J.C., Weissenbach, J., Kafatos, F.C., Collins, F.H., Hoffman, S.L.: The genome sequence of the malaria mosquito Anopheles gambiae. Science 298, 129–149 (2002)

17. Krzywinski, J., Nusskern, D.R., Kern, M.K., Besansky, N.J.: Isolation and characterization of Y chromosome sequences from the African malaria mosquito Anopheles gambiae. Genetics 166, 1291–1302 (2004)

18. Sharakhova, M.V., Xia, A., McAlister, S.I., Sharakhov, I.V.: A standard cytogenetic photomap for the mosquito Anopheles stephensi (Diptera: Culicidae): application for physical mapping. J. Med. Entomol. 43, 861–866 (2006)

19. Tsy, J.M.P., Duchemin, J.B., Marrama, L., Rabarison, P., Le Goff, G., Rajaonarivelo, V., Robert, V.: Distribution of the species of the Anopheles gambiae complex and first evidence of Anopheles merus as a malaria vector in Madagascar. Malar J. 2, 33 (2003)

20. Garcia, B.A., Caccone, A., Mathiopoulos, K.D., Powell, J.R.: Inversion monophyly in African anopheline malaria vectors. Genetics 143, 1313–1320 (1996)

21. Besansky, N.J., Krzywinski, J., Lehmann, T., Simard, F., Kern, M., Mukabayire, O., Fontenille, D., Toure, Y.T., Sagnon, N.F.: Semipermeable species boundaries between Anopheles gambiae and Anopheles arabiensis: evidence from multilocus DNA sequence variation. Proc. Natl. Acad. Sci. USA 100, 10818–10823 (2003)

Recovering True Rearrangement Events on Phylogenetic Trees

Hao Zhao and Guillaume Bourque

Genome Institute of Singapore, 138672, Republic of Singapore

Abstract. Given the gene-order of a set of contemporary genomes, the problem of recovering the rearrangement scenario that best explains these arrangements can be challenging even if the phylogeny of these species is known. Most of the existing methods can identify an optimal or near-optimal scenario in terms of parsimony but they cannot distinguish between reliable and putative events on the reconstructed tree. In this paper, we propose an efficient method to infer partial rearrangement scenarios consisting of only reliable ancestral events. Using simulations, we show that the approach allows the recovery of actual events with high sensitivity and specificity under both random and fragile rearrangement models. Finally, we also apply the approach to two real data sets.

1 Introduction

In recent years, gene order data has been intensively used to study phylogenetic trees since it provides a whole-genome view on evolution [18, 8, 6]. Because some of the simplest formulations of the problem even with 3 genomes are NP-hard [5], reconstruction algorithms have to rely on heuristics to recover a most parsimonious scenario [13, 2, 11]. Nonetheless, this did not prevent applications using the whole-genome of various vertebrate species [4, 14].

Gene-order phylogenetic reconstruction algorithms are typically evaluated based on three criteria: 1) their ability to recover the correct tree topology, 2) the total number of rearrangements in the scenario recovered [13, 2] and 3) the quality of the ancestral reconstructions [7, 3]. In the current work, we plan to evaluate these reconstructions based on a different criterion: the accuracy of the rearrangements in the recovered scenarios. To our knowledge, this has yet to be systematically analyzed. The idea here is to shift the focus from trees and ancestral reconstructions and study the quality of the inferred scenario themselves. We will only look for highly reliable (i.e. true) events as they are likely to lead to new insights in our understanding of the underlying evolutionary mechanisms. Such an analysis has seldom been performed because multiple optimal rearrangement paths frequently exist even between a pair of genomes [21]. Initially of course, this assessment will be performed on simulated data sets where the accuracy can be assessed.

Although maximum likelihood-based methods are an appealing way to try to achieve this goal, such probabilistic formulations have so far proven to be computationally prohibitive [20]. Moreover, the few developments in this area [12, 9] did not aim to estimate the accuracy of the individual ancestral events either.

G. Tesler and D. Durand (Eds.): RECOMB-CG 2007, LNBI 4751, pp. 149–161, 2007.

Our main contribution is a new approach called *Efficient Method to Recover Ancestral Events* (EMRAE) that allows the inference, on a fixed phylogenetic tree, of a partial rearrangement scenario consisting of only reliable events. As a first step, the rearrangement operations that we consider are reversals and transpositions but the method is readily expandable to other types of events. The approach relies on adjacencies shared by a significant fraction of the genomes in a given subtree. The ability to model transpositions is one of the strengths of EMRAE, since transpositions are typically harder to characterize even when only 2 genomes are considered [1, 23].

We compare EMRAE to two standard reconstruction tools: MGR [2] and GRAPPA [13] and show that EMRAE achieves comparable sensitivity but significantly higher specificity under both random and fragile models. Then, we apply our approach to two real data sets: the *Campanulaceae* Chloroplast dataset [6] and a data set consisting of 4 bacterial genomes in the *Burkholderia* family [10]. Finally, we will present some potential extensions and future directions.

2   Basic Concepts

A genome G can be represented by a signed permutation $g_1 g_2 \ldots g_n$ where each integer g_i $(1 \leq i \leq n)$ corresponds to a unique gene or marker in G. The sign of g_i represents its orientation. An *adjacency* $a(g_i, g_{i+1})$ of G is an ordered pair of integers $g_i g_{i+1}$ or its inverse $-g_{i+1} -g_i$. Denote $a \in G$ if a is an adjacency of genome G. Two adjacencies *overlap* if they share a common gene. For instance, $a(1, 2)$ and $a(-2, 3)$ overlap since they share gene 2. We view $G = g_1 g_2 \ldots g_n$ the same as its reverse $-G = -g_n -g_{n-1} \ldots -g_2 -g_1$.

For a given genome $G = g_1 g_2 \ldots g_n$, a *reversal* $r(i, j)$, where $i \leq j$, transforms G into $g_1 g_2 \ldots -g_j -g_{j-1} \ldots -g_{i+1} -g_i g_{j+1} \ldots g_n$ by reversing both the order of $g_i g_{i+1} \ldots g_{j-1} g_j$ and the sign of each gene. A *transposition* $t(i, j, k)$ on G, where $i \leq j$, exchanges the two segments $g_i \ldots g_j$ and $g_{j+1} \ldots g_k$ if $k \geq j$ or $g_k \ldots g_{i-1}$ and $g_i \ldots g_j$ if $k \leq i$. Finally, a phylogenetic tree T is a binary, unrooted tree. The leaves of T represent contemporary species while the internal nodes of T represent their ancestors.

3   Methods and Algorithms

The main idea of our approach is to identify shared adjacencies and to combine them to trace back ancestral events. For example, take the phylogenetic tree shown in Figure 1 with eight genomes $(G_1, G_2, \ldots, G_8)$ and where A, B and C represent ancestral nodes. We will now describe how to infer ancestral events on an edge $e = (A, B)$. Note that the removal of e from T partitions the genomes into two subsets $S_A = \{G_1, G_2, G_3, G_4\}$ and $S_B = \{G_5, G_6, G_7, G_8\}$.

Assume that $A = 1\ 2\ 3\ 4\ 5$ and there is only one reversal $r(2, 4)$ on e which transforms A into $B = 1\ -4\ -3\ -2\ 5$. By comparing the adjacencies of A and B, we observe that r changes two adjacencies in A, $a_1 = a(1, 2)$ and $a_2 = a(4, 5)$, and leads to two new ones in B, $b_1 = a(1, -4)$ and $b_2 = a(-2, 5)$. Other adjacencies in A are left unchanged. We say that a_1 and a_2 are the *counterparts* of b_1 and b_2 with respect to r and vice-versa. Assume that in a "perfect" scenario, a_1 and a_2 are not disrupted by any additional rearrangement event on the paths from A to every $G_i \in S_A$ and that b_1 and b_2 are not disrupted by any events on the paths from B to every $G_j \in S_B$. Then a_1 and a_2 are preserved in every genome of S_A, while neither of them can be found in a genome of S_B. We call a_1 and a_2 *conserved adjacencies* of S_A. Similarly, b_1 and b_2 are conserved adjacencies of S_B. If we can identify the conserved adjacencies in S_A and in S_B then, in a perfect scenario where these adjacencies are not reused, it will be trivial to infer the rearrangements that occur. In the example above, we would know that $a(1,2)$ and $a(4,5)$ were affected by a reversal on edge e.

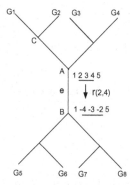

Fig. 1. Inferring an event on edge e

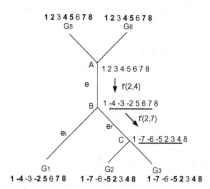

Fig. 2. Refinement step on edge e

Formally, denote by $CA(e, A)$ the sets of conserved adjacencies in S_A for an edge $e = (A, B)$:

$$CA(e, A) = \{a \mid a \in G_i, \forall G_i \in S_A \text{ and } a \notin G_j, \forall G_j \in S_B\}.$$

Let $CA(e, B)$ be defined similarly. For every edge $e \in T$, $CA(e, A)$ and $CA(e, B)$ retain the information that will be used to infer ancestral events on the edge e. We use the following *Inference Rules*:

- Reversal: Suppose we have $a_1 = a(g_{i-1}, g_i)$, $a_2 = a(g_j, g_{j+1}) \in CA(e, A)$, and $b_1 = a(g_{i-1}, -g_j), b_2 = a(-g_i, g_{j+1}) \in CA(e, B)$. We infer a reversal $r(i, j)$ from A to B.
- Transposition: If we have $a_1 = a(g_{i-1}, g_i)$, $a_2 = a(g_j, g_{j+1}), a_3 = a(g_k, g_{k+1}) \in CA(e, A)$, and $b_1 = a(g_{i-1}, g_{j+1}), b_2 = (g_k, g_i), b_3 = (g_j, g_{k+1}) \in CA(e, b)$, then we infer a transposition $t(i, j, k)$ from A to B.

In practice, we do not know the permutations of the ancestral genomes. This implies that our recovered events will be restricted to the identification of the affected sets of adjacencies and not to the specific content of the segments affected.

3.1   A Naive Algorithm: All Versus All

We implemented the above idea into a naive algorithm All_Vs_All. Specifically, to infer a reversal on an edge $e = (A, B)$, we search all the $a_1, a_2 \in CA(e, A)$ and $b_1, b_2 \in CA(e, B)$, such that they match the pairing associated with reversals described in the Inference Rules. A conserved adjacency can only be involved in a single rearrangement, so a_1, a_2 and b_1, b_2 will be removed from $CA(e, A)$ or $CA(e, B)$ once they are used. Transpositions are recovered in a similar way.

Algorithm 1. All_Vs_All $(G_1, G_2, \ldots, G_m, T)$

Input: Genomes $G_1, G_2, \ldots, G_m$, and their phylogenetic tree T
Output: Inferred events on every edge $e \in T$
1. **for** each edge $e = (A, B) \in T$ **do**
2.   Compute CA(e, A) and CA(e, B)
3. **for** each edge $e = (A, B) \in T$ **do**
4.   Infer every possible reversal r and remove the 4 related adjacencies
      from $CA(e, A)$ and $CA(e, B)$
5. **for** each edge $e = (A, B) \in T$ **do**
6.   Infer every possible transposition t and remove the 6 related adjacencies
      from $CA(e, A)$ and $CA(e, B)$

3.2   Extension: EMRAE

All_Vs_All only makes use of adjacencies that are perfectly preserved in S_A and S_B. This condition is very stringent and leaves room for possible extensions. For example, in Figure 1, if $a(1, 2)$ was disrupted by an additional rearrangement on the edge (C, G_2) then $a(1, 2)$ would not be included in $CA(e, A)$ even if it was preserved in all the other genomes of S_A (i.e. G_1, G_3 and G_4). Because of this, the reversal $r(2, 4)$ on e would be missed. In order to retain such adjacencies, we will relax our definition of conserved adjacencies.

Note that the descendants of A, S_A, can be divided into two sets, $S_{A,l}$ (left) and $S_{A,r}$ (right), such that $S_A = S_{A,l} \cup S_{A,r}$. The labelling as left or right here is arbitrary. In the example shown in Figure 1, we have that $S_{A,l} = \{G_1, G_2\}$ and $S_{A,r} = \{G_3, G_4\}$. Given $G_i \in S_{A,l}$ and $G_j \in S_{A,r}$, we denote the set of adjacencies conserved in G_i and G_j by $CA(e, A, i, j)$ and define it as:

$$CA(e, A, i, j) = \{a \mid a \in G_i, a \in G_j; a \notin G_k, \forall G_k \in S_B\}.$$

The new relaxed definition for conserved adjacencies will be:

$$CA(e, A) = \bigcup_{\{i, j \mid G_i \in S_{A,l}; G_j \in S_{A,r}\}} \{CA(e, A, i, j)\}.$$

Going back to our example, if $a(1, 2)$ was affected by an additional rearrangement on the edge (C, G_2), then $a(1, 2) \notin CA(e, A, 2, 4)$ but $a(1, 2) \in CA(e, A, 1, 4)$ such that we would still have $a(1, 2)$ in $CA(e, A)$. We note that in loosening this definition, we have insured that the conserved adjacency was observed in a pair of genomes with one genome coming from each subtree of A to ensure that it was associated with the edge e as opposed to only be associated with an internal edge in the descendants of A.

We have now relaxed our definition of conserved adjacencies but assume further that we have the more complicated example shown in Figure 2. As before, the goal is to recover the events on e and a reversal $r(2,4)$ converted the $a(1,2)$ and $a(4,5)$ in A into $a(1,-4)$, $a(-2,5)$ in B but assume that an additional reversal $r(2,7)$ affected $a(1,-4) \in B$ and $a(7,8) \in B$ and led to $a(1,-7) \in C$ and $a(4,8) \in C$. In this scenario, $\{a(1,2), a(4,5)\} \subseteq CA(e,A)$ and $a(-2,5) \in CA(e,B)$ but $a(1,-4)$ would be "missing" in $CA(e,B)$ since it can not be found in the right subtree of B. Because of this, the adjacency would actually have shifted to the left subtree of B. In our second extension to All_Vs_All we will seek to recover such adjacencies by adding them to the correct edges.

For convenience, in Figure 2, we label the edges leading to the left and right subtree of B as e_l and e_r respectively. We say that an adjacency $a \in CA(e_l, G_1)$ is *isolated* if it does not overlap with any adjacency in $CA(e_l, B)$. For instance, $a(1,-4) \in CA(e_l, G_1)$ is isolated. From Figure 2, we observe that on the edge e_r both $a(1,-7)$ and $a(4,8)$ which are in $CA(e_r, C)$ overlap with $a(7,8) \in CA(e_r, B)$. This suggests that $a(1,-7)$ and $a(4,8)$ are involved in an event that took place on e_r but that one of their counterpart adjacencies is missing in $CA(e_r, B)$. This missing adjacency should include the genes 1 and 4 although their orientation is undetermined at this point. Because the isolation of $a(1,-4) \in CA(e_l, G_1)$ appears to come from the disruption of $a(1,-4)$ on e_r, we want to add it into $CA(e,B)$.

Formally, given an edge $e = (A,B)$, its left branch $e_l = (B,C)$ and its right branch $e_r = (B,D)$, we define the *refinement* of $CA(e,B)$ as follows. If $a(i,j) \in CA(e_l, C)$ is isolated, and if $a(i,k)$ and $a(j,m)$ in $CA(e_r, D)$ overlap with $a(k,m) \in CA(e_r, B)$, then add $a(i,j)$ into $CA(e,B)$. Similarly, add the isolated adjacencies in $CA(e_r, D)$ that satisfy the reciprocal condition on e_l back into $CA(e,B)$. Currently, our refinement is only based on using relaxed conserved adjacencies and we do no use those newly-added adjacencies to refine other edges. This implies that the refinement step does not depend on the order of traversal of the edges.

Together, these two extensions form the Efficient Method to Recover Ancestral Events (EMRAE).

4 Simulations

4.1 Random Breakage Model

We generated simulated data sets with m genomes containing n genes each. First, a random tree T is generated. Next, one internal node of T is identified as the identity permutation and we mimic the evolutionary history by performing k rearrangements on each edge e. This k is a random integer such that $0 \leq k \leq 2 * \mu$, where μ is the *evolutionary rate* of T and corresponds to the average

---

Algorithm 2. EMRAE $(G_1, G_2, \ldots, G_m, T)$

Input: Genomes $G_1, G_2, \ldots, G_m$, and their phylogenetic tree T
Output: Inferred events on every edge $e \in T$
1. **for** each edge $e = (A, B) \in T$ **do**
2.   Compute $CA(e, A)$ and $CA(e, B)$
3. **for** each edge $e = (A, B) \in T$ **do**
4.   Refine $CA(e, A)$ and $CA(e, B)$
6. **for** each edge $e = (A, B) \in T$ **do**
7.   Infer every possible reversal r and remove the 4 related adjacencies
      from $CA(e, A)$ and $CA(e, B)$
8. **for** each edge $e = (A, B) \in T$ **do**
9.   Infer every possible transposition t and remove the 6 related adjacencies
      from $CA(e, A)$ and $CA(e, B)$

---

number of events on each edge. This procedure is repeated down to every leaf node. Finally we use the tree topology and the obtained permutations as the input to the different algorithms. The positions in the permutations affected by rearrangement events are randomly selected according to a uniform distribution.

We set $m = 7, n = 100$ and tested 3 distinct models: reversal-only, equally likely reversal and transposition, and transposition-only. We evaluated the performance of All_Vs_All and EMRAE by comparing them to MGR [2] and GRAPPA [13]. As discussed in Section 3, an event recovered by All_Vs_All or EMRAE only indicates a set of affected adjacencies. Although MGR and GRAPPA reconstruct trees and ancestral genomes, they do not directly provide a detailed rearrangement scenario as part of their output. As a surrogate, we used GRIMM [22] to produce a most parsimonious scenario on each edge of the trees recovered by MGR and GRAPPA. Moreover, although transpositions are not directly considered by MGR or GRAPPA, they can be mimicked by 3 consecutive reversals. For instance, suppose a transposition $t(2, 3, 5)$ transforms a permutation $A = 1\ 2\ 3\ 4\ 5\ 6$ into $B = 1\ 4\ 5\ 2\ 3\ 6$. A possible way to mimic the transposition t is to first perform $r(2, 3)$ to transform A into $A_1 = 1\ -3\ -2\ 4\ 5\ 6$. Then perform $r(4, 5)$ on A_1 to get $A_2 = 1\ -3\ -2\ -5\ -4\ 6$. And finally, perform $r(2, 5)$ on A_2 to get $B = 1\ 4\ 5\ 2\ 3\ 6$. There is a total of 6 possible ways to mimic a transposition with a sequence of 3 reversals. For every edge of a tree produced by MGR or GRAPPA, we will process the rearrangement scenario and look for such triplets of reversals and label them as putative transpositions.

For a given data set and a list of events inferred by one of the algorithms, we define:

$$Sensitivity = \frac{\text{Nb inferred real events}}{\text{Total nb of real events}} \times 100,$$

$$Specificity = \frac{\text{Nb inferred real events}}{\text{Total nb of inferred events}} \times 100.$$

The sensitivity measures the proportion of real events that are recovered while the specificity measures the proportion of the predictions that are correct.

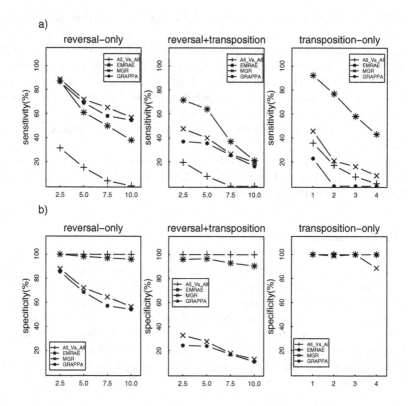

Fig. 3. Performance of the different methods: a) sensitivities, b) specificities. The x-axis is the evolutionary rate μ.

We generated 30 simulation instances for different μ and use the average sensitivity and specificity to evaluate the performance of the different algorithms. The results are shown in Figure 3 (see also Table S1 in the Appendix). Note that we used lower values of μ in the transposition-only model since transpositions are harder to recover and are assumed to be more rare in many actual data sets.

Our first observation based on Figure 3 is that EMRAE has a much higher sensitivity than All_Vs_All. This indicates that the improvements we made over All_Vs_All were significant. Also, EMRAE has a higher sensitivity as compared to MGR and GRAPPA in all models that include transpositions. Interestingly, we observe that although the sensitivity of MGR and GRAPPA in the reversal-only model is higher than that of EMRAE, this difference is marginal. This is somewhat unexpected given that this is the optimal model for both MGR and GRAPPA.

Moving on to the specificity, we note that both All_Vs_All and EMRAE have scores that are consistently above 90%. In the two models involving reversals, the specificity is significantly higher than MGR and GRAPPA. This is one of the key features of our approach and confirms that the conservative inference leads to predictions that are almost always correct. Note that the specificity of MGR and GRAPPA for the transposition-only model is high because they make very few predictions.

To test if the order of inference of reversals and transpositions could affect the output, we also tried running EMRAE while reversing step 6 and 7 in algorithm 2. Overall, we found that the sensitivities and specificities were consistent with the results shown in Figure 3. This is somewhat expected given that if a conserved adjacency was associated with both a reversal and a transposition, then at least one of the predicted events would be wrong but recall that EMRAE has a very high specificity.

In summary, EMRAE achieves significant sensitivity and comparable to both MGR and GRAPPA. In terms of specificity, EMRAE clearly outperforms the other two and leads to the identification of true events.

4.2   Fragile Regions Model

There has been a heated debate over the existence in genomes of *fragile* regions that are more prone to breakage [16, 19, 15, 17]. We were also interested in measuring the impact of enforcing a certain degree of breakpoint reuse in our simulations. Specifically, we divide the positions $(g_i, g_i + 1)$ into *weak* and *strong* ones. Weak positions are fragile and are more likely to be break. In our model this is achieved with two parameters: x the proportion of weak positions and y how much "weaker" the weak positions are as compared to the strong ones. That is, for a position i:

$$y = \frac{Pr(i \text{ breaks} \mid i \text{ is weak})}{Pr(i \text{ breaks} \mid i \text{ is strong})}.$$

We increase or decrease the breakpoint reuse rate by adjusting the pair (x, y).

To make our simulations more realistic, we purposely selected the parameters to produce scenarios that would be comparable to the *Campanulaceae* data set [6]. Specifically, we chose $m = 13, n = 107$. To measure the extent of breakpoint reuse, we relied on the notion of *strip* which is a maximal segment of genes in the same order in all genomes. The 107 genes in the *Campanulaceae* data set can be compressed into 36 strips. Since inputting this data set into MGR returns a scenario with 65 reversals [2] and given that this tree has 23 edges, this implies an evolutionary rate $\mu \approx 2.82$. We note however that simulating a tree under the random breakage model described in the previous section with m=13, n=107 and an evolutionary rate $\mu = 3$ (which implies a total of approximately $23 \times 3 = 69$ reversals), leads to a data set that contains on average 82 strips. This is much higher than the 36 observed in the *Campanulaceae* data set and suggests significant breakpoint reuse. We will repeat the simulations but adjust the parameters x, y and μ until two conditions are met: 1) the score of the tree, as inferred by MGR, is ≈ 65 and 2) the number of strips is ≈ 36. Finally, results for different values of x and y and $\mu = 3.5$ are shown in Table 1. The values $(x, y) = (0.06, 20)$ appear to best match the *Campanulaceae* data set.

Table 1. Breakpoint reuse and the number of strips in the permutations. The number of strips gets smaller as the breakpoint reuse rate increases.

x	y	#MGR	#strips
0.1	5	81.3	80.6
0.1	10	75	52
0.06	15	66.7	45
0.06	20	64.8	34

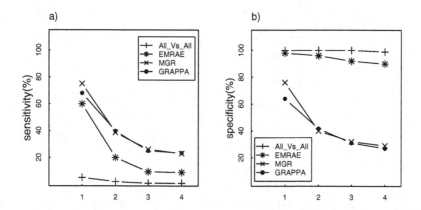

Fig. 4. Performance of the different algorithms on the fragile reversal-only model: a) sensitivities, b) specificities. The four data points on the x-axis correspond to $(x, y) = (0.1, 5), (0.1, 10), (0.06, 15)$, and $(0.06, 20)$, respectively.

Under these settings and a reversal-only model, we compared the predictions EMRAE, MGR and GRAPPA (see Figure 4 and also Table S1 in the Appendix). As expected, the sensitivity of all approaches is affected by breakpoint reuse but we note that the specificity of EMRAE remains exceedingly high.

5   Result on Real Data

We have already presented the *Campanulaceae* Chloroplast data set [6]. The result of applying EMRAE to this data set is shown in Figure 5a (note that we have used as additional input MGR's predicted topology). EMRAE only inferred 9 events (5 reversals and 4 transpositions), but recall that this is consistent with what we have observed in the simulations that included a high rate of breakpoint reuse (Figure 4a). Relying on the same simulations, we expect most of these 9 events to be true evolutionary events (Figure 4b).

We were interested in measuring the overlap between the predictions of EMRAE and MGR to see if the former was only a subset of the latter. Recall that MGR returned a scenario with 65 reversals. Analyzing this scenario suggested

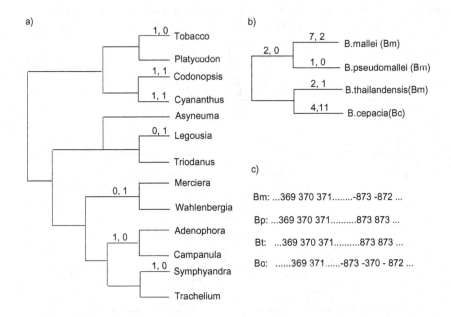

Fig. 5. EMRAE's predictions on a) the Campanulaceae Chloroplast dataset, b) Burkholdria data. The two numbers on an edge (if any) are the number of reversals and transpositions inferred on that edge. c) Parts of the permutations associated with Bm, Bp, Bt and Bc associated with one of the transposition recovered on the Bc lineage.

that 3 of these reversals actually corresponded to a transposition. Overall, we found that the two scenarios shared 3 common reversals and 1 transposition confirming that EMRAE recovered distinct high quality evolutionary events (2 reversals and 3 transpositions).

The second data set that we analyzed consists of four bacterial species in the *Burkholderia* family [10]. These genomes share 2435 genes spread on two chromosomes. Very few genes are exchanged between chromosomes in this phylogeny and those genes are omitted for simplicity. Lin et al. [10] reported a scenario with 240 reversals, of which 3 could be associated to a single transposition, leading to a scenario with 238 events. The result of applying EMRAE using the known topology is shown in Figure 5b. The algorithm predicts 30 events: 16 reversals and 14 transpositions. For this data set most (14 out of 16) of EMRAE's reversals were also found in MGR's predictions. That said, these 14 common high-quality reversals were buried in a very large set of predictions in the MGR scenario. The only transposition predicted by MGR was also found by EMRAE. Overall, our algorithm predicted 13 additional transpositions but this is to be expected based on Figure 3. An example of such a transposition is shown in Figure 5c.

Because in both real data sets the number of predictions is much smaller than the total number of rearrangements predicted by MGR, we deduce that breakpoint reuse is common in these scenarios. Nevertheless, we remain confident that the predictions that are made by EMRAE are of high quality based on our simulations.

6   Discussion

We have presented an efficient method to infer partial rearrangement scenarios consisting of only reliable rearrangement events.

Some of the important future directions include:

- Extensions to predict other types of events: 1) translocations, fusions and fissions for multi-chromosomal genomes and 2) insertions and deletions to expand to genomes with unequal content. This would require the modelling of the impact of these new events on adjacencies and should be relatively straightforward.
- Application to larger genomes (e.g. sequenced mammalian genomes) followed by an in-depth analysis of the recovered events to gain mechanistic insights into the likely causes of rearrangements events.

Improving our understanding of the evolutionary forces driving large-scale rearrangement events has been a promise only partially fulfilled by previous computational analysis. Perhaps trusting only highly reliable ancestral events, such as the ones obtained using this approach, will lead to new insights. Ultimately, this knowledge will feed back into the design of more accurate rearrangement models and scenarios.

Acknowledgements

We would like to thank the reviewers for helpful suggestions. This work is supported by funds from the Biomedical Research Council (BMRC) of Singapore.

References

[1] Bafna, V., Pevzner, P.A.: Sorting by transpositions. SIAM Journal on Discrete Mathematics 11(2), 224–240 (1998)
[2] Bourque, G., Pevzner, P.A.: Genome-scale evolution: reconstructing gene orders in the ancestral species. Genome Res. 12, 26–36 (2002)
[3] Bourque, G., Tesler, G., Pevzner, P.A: The convergence of cytogenetics and rearrangement-based models for ancestral genome reconstruction. Genome Res. 16(3), 311–313 (2006)
[4] Bourque, G., Zdobnov, E.M., Bork, P., Pevzner, P.A., Tesler, G.: Comparative architectures of mammalian and chicken genomes reveal highly variable rates of genomic rearrangements across different lineages. Genome Res. 15(1), 98–110 (2005)

[5] Caprara, A.: Formulations and complexity of multiple sorting by reversals. In: RECOMB 99, pp. 84–93 (1999)

[6] Cosner, M., Jansen, R., Moret, B., Raubeson, L., Wang, L., Warnow, T., Wyman, S.: A new fast heuristic for computing the breakpoint phylogeny and experimental phylogenetic analyses of real and synthetic data. In: ISMB00, pp. 104–115 (2000)

[7] Froenicke, L., Caldes, M.G., Graphodatsky, A., Muller, S., Lyons, L.A, Robinson, T.J, Volleth, M., Yang, F., Wienberg, J.: Are molecular cytogenetics and bioinformatics suggesting diverging models of ancestral mammalian genomes?. Genome Res. 16(3), 306–310 (2006)

[8] Hannenhalli, S., Chappey, C., Koonin, E., Pevzner, P.: Genome sequence comparison and scenarios for gene rearrangements: A test case. Genomics 30, 299–311 (1995)

[9] Larget, B., Simon, D., Kadane, J., Sweet, D.: A bayesian analysis of metazoan mitochondrial genome arrangements. Mol. Biol. Evol. 22(3), 486–495 (2005)

[10] Lin, C.H., Bourque, G., Tan, P.: Comparative analysis of Burkholderia species reveals an association between large-scale and fine-scale divergence in prokaryotes. In: preparation (2007)

[11] Ma, J., Zhang, L., Suh, B.B., Raney, B.J., Burhans, R.C, Kent, W J., Blanchette, M., Haussler, D., Miller, W.: Reconstructing contiguous regions of an ancestral genome. Genome Res. 16(12), 1557–1565 (2006)

[12] Miklos, I.: MCMC genome rearrangement. Bioinformatics 19(90002), 130ii–137 (2003)

[13] Moret, B.M.E., Wyman, S., Bader, D.A., Warnow, T., Yan, M.: A new implementation and detailed study of breakpoint analysis. In: PSB 2001, pp. 583–594 (2001)

[14] Murphy, W.J., Larkin, D.M., van der Wind, A.E.-, Bourque, G.: Dynamics of mammalian chromosome evolution inferred from multispecies comparative maps. Science 309(5734), 613–617 (2005)

[15] Peng, Q., Pevzner, P.A, Tesler, G.: The fragile breakage versus random breakage models of chromosome evolution. PLoS Comput. Biol. 2(2), 14 (2006)

[16] Pevzner, P., Tesler, G.: Human and mouse genomic sequences reveal extensive breakpoint reuse in mammalian evolution. Proc. Natl. Acad. Sci. U S A 100(13), 7672–7677 (2003)

[17] Sankoff, D.: The signal in the genomes. PLoS Comput. Biol. 2(4), e35 (2006)

[18] Sankoff, D., Leduc, G., Antoine, N., Paquin, B., Lang, B., Cedergren, R.: Gene order comparisons for phylogenetic inference: Evolution of the mitochondrial genome. Proceedings of the National Academy of Sciences USA 89, 6575–6579 (1992)

[19] Sankoff, D., Trinh, P.: Chromosomal breakpoint reuse in genome sequence rearrangement. J. Comput. Biol. 12(6), 812–821 (2005)

[20] Savva, G., Dicks, J.L., Roberts, I.N.: Current approaches to whole genome phylogenetic analysis. Briefings in Bioinformatics 4(1), 63–74 (2003)

[21] Siepel, A.: An algorithm to find all sorting reversals. In: RECOMB02, pp. 281–290 (2002)

[22] Tesler, G.: GRIMM: genome rearrangements web server. Bioinformatics 18(3), 492–493 (2002)

[23] Walter, M., Dias, Z., Meidanis, J.: A new approach for approximating the transposition distance. In: De La Fuenta, P. (ed.) SPIRE, pp. 199–208. IEEE Computer Society Press, Los Alamitos (2000)

Appendix

Table S1. Performance of the different methods in terms of the number of predicted events: a) number of true predictions, and b) number of false predictions. Each number in the last column corresponds to the average number of real events on a tree. The $\mu = 1, 2, 3, 4$ for reversal-only (fragile) correspond to $(x, y) = (0.1, 5), (0.1, 10), (0.06, 15), (0.06, 20)$, respectively.

a)

model	μ	All_Vs_All	EMRAE	MGR	GRAPPA	nb of real events
reversal-only (random)	2.5	8.6	24	24.4	23.8	27.5
	5	8.4	33.6	39.5	38	55
	7.5	3.5	41.3	53.8	48	82.5
	10	0	41.8	62.7	60.2	110
reversal + transposition (random)	2.5	5.4	19.7	13	10.2	27.5
	5	4.5	35.2	22	19.5	55
	7.5	0	30.5	22	21.2	82.5
	10	0	23.3	21.6	18.2	110
transposition-only (random)	1	3.9	10.1	5	2.5	11
	2	3.8	16.9	4.6	0	22
	3	2.5	19.1	5.3	0	33
	4	0.9	18.9	3.8	0	44
reversal-only (fragile)	1	4	48.6	60.7	55	80.5
	2	1.7	16.2	31.4	32.2	80.5
	3	0.8	7.6	21	20.2	80.5
	4	0.8	7.3	18.6	18.9	80.5

b)

model	μ	All_Vs_All	EMRAE	MGR	GRAPPA	nb of real events
reversal-only (random)	2.5	0	0	3.3	4	27.5
	5	0	0.6	15.4	17.5	55
	7.5	0	1.3	29.66	35.9	82.5
	10	0	1.7	48.3	50.8	110
reversal + transposition (random)	2.5	0	0.2	26.9	31.5	27.5
	5	0	0.7	57.4	62.2	55
	7.5	0	1.2	102.4	104.3	82.5
	10	0	1.2	144.3	149	110
transposition-only (random)	1	0	0	0	0	11
	2	0	0	0	0	22
	3	0	0	0	0	33
	4	0	0	0	0	44
reversal-only (fragile)	1	0	1	19.1	30.9	80.5
	2	0	0.7	47.1	44.8	80.5
	3	0	0.7	44.8	45.1	80.5
	4	0	0.8	45.6	47.5	80.5

Parts of the Problem of Polyploids in Rearrangement Phylogeny

Chunfang Zheng, Qian Zhu, and David Sankoff

Departments of Biology, Biochemistry, and Mathematics and Statistics,
University of Ottawa, Ottawa, Canada K1N 6N5
{czhen033,qzhu012,sankoff}@uottawa.ca

Abstract. Genome doubling simultaneously doubles all genetic markers. Genome rearrangement phylogenetics requires that all genomes analyzed have the same set of orthologs, so that it is not possible to include doubled and unduplicated genomes in the same phylogeny. A framework for solving this difficulty requires separating out various possible local configurations of doubled and unduplicated genomes in a given phylogeny, each of which requires a different strategy for integrating genomic distance, halving and rearrangement median algorithms. In this paper we focus on the two cases where doubling precedes a speciation event and where it occurs independently in both lineages initiated by a speciation event. We apply these to a new data set containing markers that are ancient duplicates in two yeast genomes.

1 Introduction

Basic rearrangement phylogeny methods require that the genomic content be the same in all the organisms being compared, so that every marker (whether gene, anchor, probe binding site or chromosomal segment) in one genome be identified with a single orthologous counterpart in each of the others, though adjustments can be made for a limited amount of marker deletion, insertion and duplication.

Many genomes have been shown to result from an ancestral doubling of the genome, so that every chromosome, and hence every marker, in the entire genome is duplicated simultaneously. Subsequently, the doubled genome evolves through mutation at the DNA sequence level and by chromosomal rearrangement, through intra- and interchromosomal movement of genetic material. This movement can scramble the order of markers, so that the chromosomal neighbourhood of a marker need bear no resemblance to that of its duplicate.

The present-day genome, which we refer to here as a doubling descendant, can be decomposed into a set of duplicate or near-duplicate markers dispersed among the chromosomes. There is no direct way of partitioning the markers into two sets according to which ones were together in the same half of the original doubled genome. Genomic distance or rearrangement phylogeny algorithms are not applicable to doubling descendants, since there is a two-to-one relationship between markers in the doubling descendant and related species whose divergence predates the doubling event, whereas these algorithms require a one-to-one correspondence.

G. Tesler and D. Durand (Eds.): RECOMB-CG 2007, LNBI 4751, pp. 162–176, 2007.

We have undertaken a program [11,9] of studying rearrangement phylogeny where doubling descendants are considered along with related unduplicated genomes. We believe there is no other computationally-oriented literature on this particular problem. To focus on the problem of marker ambiguity in doubling descendants, and to disentangle it from the difficulties of constructing phylogenies, we pose our computational problems only within the framework of the "small" phylogenetic problem, i.e., identifying the ancestral genomes for a given phylogeny that jointly minimize the sum of the rearrangement distances along its branches.

In Section 2, we outline a model for generating an arbitrary pattern of doubled descendants observed at the tips of a given phylogeny. Based on this model, we then present an simple algorithm for inferring the doubling status of the ancestral genomes in terms of an economical set of doubling events along the branches of the phylogeny. Once we have the ancestral doubling statuses, we can approach the actual rearrangement problem.

First, in Section 3, we identify three kinds of component of this problem for which algorithms already exist, one a calculation of the genomic distance between two given genomes with clearly identified orthologs, i.e., the minimum number of rearrangements necessary to transform one genome into another; the second a "halving" algorithm for inferring the genome of a doubled genome based on internal evidence from its modern descendant only, and the third a "medianizing" process for inferring an ancestral genome from its three neighbouring genomes in a binary branching tree.

In Section 4, we discuss our recent papers [11,9] on incorporating algorithms for the three components into an overall procedure for inferring ancestral genomes in the case of one doubling descendant and two related unduplicated genomes. The contribution of the present paper starts in Section 5 where we analyze two ways of relating genomes from two doubling descendants, one where they result from a single genome doubling event followed by a speciation, and the other where speciation precedes two genome doublings, one in each lineage. In Section 7, we apply these two methods to a large data set on yeast.

1.1   Terminology and Scope

In biology, the concept of genome doubling is usually expressed as tetraploidization or autotetraploidization, and the both the doubled genome and its doubling descendant are called tetraploid, even though, generally, the descendants soon undergo a process called (re-)diploidization and function as normal diploids, still carrying a full complement of duplicate markers that evolve independently of each other. Though unambiguous in biological context, implicit in this terminology are many assumptions that are not pertinent to our study. In the yeast data we study here, for example, *Saccharomyces cerevisiae* exists during most of its life cycle as a haploid, only sometimes as a diploid, while *Candida glabrata* exists uniquely as a haploid.

In our considerations, the key aspect of genome doubling is the global duplication of chromosomes and markers at the moment of doubling. Ploidy is not relevant in that in any organism that reproduces by meiosis or even by mitosis, the order of the markers on any of the haploid components (e.g., maternal versus paternal chromosomes) is essentially identical. There may be different alleles, or other local differences, but the order is basically invariant. Ongoing variation and evolution at the level of chromosomal structure in an individual or species are considered negligible in comparison with the major rearrangements that exist between genomes separated on an evolutionary time scale.

Although this paper is about polyploidy, then, we will rely largely on terminology independent of ploidy: genome doubling, doubling descendant, unduplicated genomes, genome halving.

The marker complement of a genome may also double by another process, allotetraploidization, or fusion of two different genomes, a kind of hybridization that is probably at least as important biologically as the doubling of a single genome we focus on in this paper. We do not consider this process here, for three reasons. One is our interest in exploring the essential difficulty in the mathematics of doubling, namely the complete ambiguity as to which set of duplicate markers were together in each of the two copies of the original genome. For hybrid doubled genomes, DNA sequence evidence from related but unduplicated genomes can generally resolve this ambiguity [5]. Second, hybrids require reticulate phylogenies which, though of interest themselves, constitute an unwanted layer of difficulty that we wish to keep separate at this stage. Finally, some of the most interesting doubling events (outside the plant kingdom), such as the ones hypothesized in the "2R" model of early vertebrate evolution or the well-established doubling in the ancestor of budding yeasts *Saccharomyces cerevisiae* and *Candida glabrata*, which furnish the empirical example for this paper, are usually treated as doubling of a single genome.

2   Generation and Inference of Polyploidy

Our algorithms require genomic sequence data or other high resolution marker data spanning the entire genome. This, of course, is only available in a limited number of phylogenetic domains within the eukaryotes, and then only from selected organisms. Our analysis may also benefit from information on doubling status not only about the sequenced or mapped genomes, but also from closely related organisms. Fortunately such information is much easier to obtain experimentally and to come by in the literature, though ancestral events often require inferential leaps based on the number of chromosomes or the distribution of the number of copies of each marker.

Our first task, given some mixture of doubling descendants and unduplicated genomes related by a phylogenetic tree, is to infer the doubling status of the all the ancestral genomes. Under the simplifying assumptions that all ploidies are powers of two and can only remain unchanged or change by a factor of two

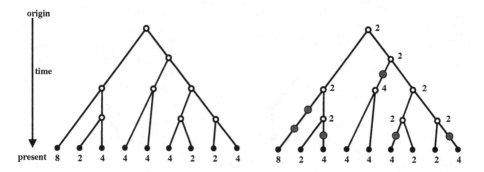

Fig. 1. Example of doubling inference problem. Genomes observed only for leaves (filled dots) of phylogeny. 2 = diploid unduplicated genome. Inferred doubling events indicated by red dots.

at each step, and the parsimony criterion that the number of doubling steps is to be minimized, the task is achieved by the recurrence

$$\Pi(v) = \min_{\text{daughter species } u \text{ of } v} \Pi(u)$$

at each ancestral vertex v of a phylogenetic tree, as depicted in Fig. 1.

Once Π is inferred, the doubling events may be inferred to occur on those branches of the tree where the Π differs at the two ends. This is also depicted in Fig.1. In the ensuing sections, we will illustrate the local configurations giving rise to various inference problems by highlighting appropriate portions of the tree in Fig. 1.

3   Existing Resources

Once we have inferred the doubling status of the ancestral genomes, how are we to approach our original problem: to reconstruct the marker order of the ancestral genomes and thus infer the cost of the phylogeny in terms of rearrangement events? Here we discuss some basic elements of the solution.

Genomic distance. Distance based on genomic structure $d(X, Y)$ is calculated by linear-time rearrangement algorithms for finding the minimum number of operations necessary to convert one genome X into another Y. Genomic distance is defined only between genomes of the same ploidy, as highlighted in the leftmost example depicted in Fig. 2.

The biologically-motivated rearrangement operations we consider include inversions (implying as well change of orientation) of chromosomal segments containing one or more markers, reciprocal translocations (of telomere-containing segments – suffixes or prefixes – of two chromosomes) and chromosome fission or fusion. We use the versatile rearrangement algorithm of Bergeron *et al.* [1], which we constrain to allow only the operations we have listed.

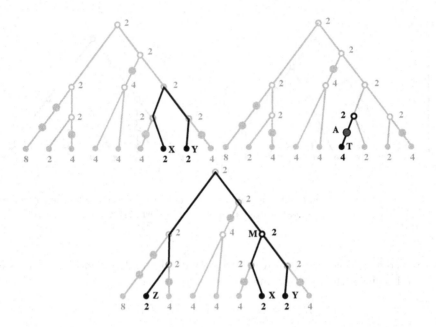

Fig. 2. Clockwise, from upper left: Genomic distance, Genome halving, Rearrangement median

Genome halving. Given a genome T containing a set of markers, each of which appears twice on the genome, on the same or on different chromosomes, how can we construct a genome A containing only one copy of each marker, and such that the genome $A \oplus A$ consisting of two copies of each chromosome in A minimizes $d(T, A \oplus A)$? This problem is illustrated in the rightmost example in Fig. 2. Here we use a linear-time algorithm for solving this problem [6].

Rearrangement median. Given three genomes X, Y and Z, how can we find the *median* genome M such that $d(X, M) + d(Y, M) + d(Z, M)$ is minimized. For this NP-hard problem, illustrated in the bottom example in Fig. 2. we implement a heuristic using the principles of Bourque's MGR [2], but based on the constrained version of the Bergeron *et al.* [1] algorithm.

4   Parts Already in Place

In this section we discuss heuristics for prototypical phylogeny problems involving doubling descendant, and either one or two related unduplicated genomes.

Let T be a doubling descendant, i.e., with n different chromosomes, and $2m$ markers, $g_{1,1} \cdots, g_{1,m}; g_{2,1}, \cdots, g_{2,m}$, dispersed in any order on these chromosomes. For each i, we call $g_{1,i}$ and $g_{2,i}$ "duplicates", and the subscript "1" or "2" is assigned arbitrarily. A potential ancestral doubled genome of T is written $A \oplus A$, and consists of $2n'$ chromosomes, where some half (n') of the chromosomes

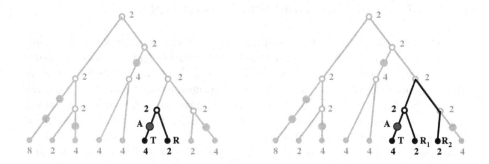

Fig. 3. Genome halving with one (left) or two (right) unduplicated outgroups

contains exactly one of each of $g_{1,i}$ or $g_{2,i}$ for each $i = 1, \cdots, m$. The remaining n' chromosomes are each identical to one in the first half, in that where $g_{1,i}$ appears on a chromosome in the first half, $g_{2,i}$ appears on the corresponding chromosome in the second half, and *vice versa*. We define A to be either of the two halves of $A \oplus A$, where the subscript 1 or 2 is suppressed from each $g_{1,i}$ or $g_{2,i}$. These n' chromosomes, and the m markers they contain, $g_1, \cdots, g_m$, constitute a potential ancestor of T that incurred the doubling event .

Genome halving with an outgroup. With reference to the left of Fig. 3, consider T and and a related unduplicated genome R with markers orthologous to $g_1, \cdots, g_m$. Our problem is to find an unduplicated genome A that minimizes

$$D(T, R) = d(R, A) + d(A \oplus A, T). \tag{1}$$

Our solution in [11], as on the left of Fig. 4, is to generate the set $\mathbf{S}$ of genome halving solutions, then to focus of the subset $X \in \mathbf{S'} \subset \mathbf{S}$ where $d(R, X)$ is minimized. We then minimize $D(T, R)$ by seeking heuristically for A along any trajectory between elements of $\mathbf{S'}$ and the outgroups.

Genome halving with two outgroups. With reference to the right of Fig. 3, consider T and two unduplicated genomes R_1 and R_2 with markers orthologous to $g_1, \cdots, g_m$. Our problem here is to find a diploid genome A and a median genome M of A, R_1 and R_2 that minimize

$$D(T, R_1, R_2) = d(R_1, M) + d(R_2, M) + d(A, M) + d(A \oplus A, T). \tag{2}$$

Our solution in [9], as on the right of Fig. 4, is to generate the set $\mathbf{S}$ of solutions of the genome halving problem, then to focus of the subset $X \in \mathbf{S'} \subset \mathbf{S}$ where $d(R_1, M) + d(R_2, M) + d(X, M)$ is minimized. Then the A minimizing $D(T, R_1, R_2)$ is sought, heuristically, along all trajectories between all elements $X \in \mathbf{S'}$ and $M(X)$.

5   The Case of Two Doubling Descendants

Two related doubled descendants may arise in two ways, depending on the timing of the speciation event in relation to the doubling. Either speciation at V follows a single doubling event, as at A on the left of Fig. 5, or the speciation precedes two independent doubling events in the two lineages, as at A and B on the right of the figure. Knowing which of the two scenarios is correct depends on knowing whether their common ancestor is doubled or not, information obtained from the algorithm in Section 2 or other data.

We will introduce new methods based on tweaking the distance and halving algorithms, conserving the optimality of the solutions, but allowing one of them to affect the arbitrary choices required to construct the solution for the other. First we sketch the halving algorithm.

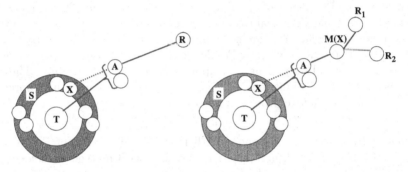

Fig. 4. Halving a doubling descendent T, with one (R) or two (R_1, R_2) unduplicated outgroups. The double circles represent two copies of potential ancestral genomes, including solutions to the genome halving in **S**, and those on best trajectories between **S** and outgroups.

5.1   Halving

Without entering into all its details, we can present enough of the essentials of the halving algorithm to understand the techniques we use in our heuristics.

As a first step each marker x in a doubled descendant is replaced by a pair of vertices (x_t, x_h) or (x_h, x_t) depending if the DNA is read from left to right or right to left. The duplicate of marker $x = (x_t, x_h)$ is written $\bar{x} = (\bar{x}_t, \bar{x}_h)$. Of course $\bar{\bar{a}} = a$.

Following this, for each pair of neighbouring markers, say (x_t, x_h) and (y_h, y_t), the two adjacent vertices x_h and y_h are linked by a black edge, denoted $\{x_h, y_h\}$ in the notation of [1]. For a vertex at the end of a chromosome, say y_t, it generates a virtual edge of form $\{y_t, O\}$.

The edges thus constructed are then partitioned into *natural graphs* according to the following principle: If an edge $\{a, b\}$ belongs to a natural graph, then so does some edge of form $\{\bar{a}, c\}$ and some edge of form $\{\bar{b}, d\}$. If a natural graph has an even number of edges, it can be shown that in all optimal ancestral

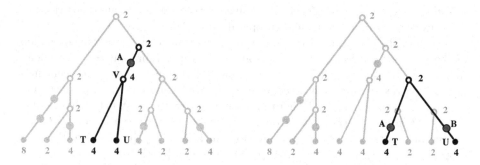

Fig. 5. Left: Doubling, then speciation. Right: Speciation, then two independent doublings.

doubled genomes, the edges coloured gray, say, representing adjacent vertices in the ancestor, and incident to one of the vertices in this natural graph, necessarily have as their other endpoint another vertex within the same natural graph[1].

For all other natural graphs, there are one or more ways of grouping them pairwise into *supernatural graphs* so that an optimal doubled ancestor exists such that the edges coloured gray incident to any of the vertices in a supernatural graph have as their other endpoint another vertex within the same supernatural graph.

Along with the multiplicity of solutions caused by different possible constructions of supernatural graphs, within such graphs and within the natural graphs, there may be many ways of drawing the gray edges. Without repeating here the lengthy details of the halving algorithm, it suffices to note that these alternate ways can be generated by choosing one of the vertices within each supernatural graph as a starting point.

5.2 Doubling First

Given two doubling descendants T and U as on the left of Figure 5, we would ideally like to find the doubling descendant V that minimizes $d(T,V)+d(V,U)+d(V,A\oplus A)$, where A is any solution of the halving problem on V. Though d is calculated in linear time, multiple genome rearrangement problems based on d (e.g., the median problem in Section 3) are hard, so here we propose a somewhat constrained version of our problem, where V is assumed to be on a shortest trajectory between T and U. Because $d(T,V)+d(V,U)=d(T,U)$ is then constant, the problem becomes that of finding V to minimize $d(V,A\oplus A)$.

Because it is an edit distance, a genomic distance measurement $d(T,U)$ is associated with at least one trajectory containing $d(T,U)-1$ genomes as well as

---

[1] Space precludes us from elaborating on the connection between the optimality criterion – the minimum number of rearrangements to transform the doubled ancestor to the doubled descendant – and the nature of the bicoloured graph defined by the black and gray edges. Suffice it to indicate that this involves maximizing the number of (alternating coloured) cycles and certain paths that make up this graph.

T and U themselves, where each successive pair of genomes along the trajectory differ by exactly one rearrangement operation.

Before explaining a heuristic search for a solution to the constrained version of the problem, we recall the edge notation we use to represent the adjacencies in a genome [1]. If two vertices a and b from different markers are adjacent in a genome, we represent this by an edge $\{a, b\} = \{b, a\}$; for a vertex c is at the end of a chromosome and hence adjacent to no other vertex, we construct a virtual edge $\{c, O\}$. Then any rearrangement operation can be represented by an operation on one or two terms in the representation, such as $\{a, b\}, \{c, d\} \to \{b, d\}, \{a, c\}$ or $\{a, b\} \to \{b, O\}, \{a, O\}$ or $\{a, b\}, \{c, O\} \to \{b, O\}, \{a, c\}$.

We initialize $T^* = T, U^* = U$. Then our heuristic consists of a search, at each step, for the "most promising" operation that moves T^* towards U^* or U^* towards T^*. For each operation, we define a score $W = x + 6y$ as follows. The y component, which is heavily weighted, measures whether the operation actually diminishes $d(V, A \oplus A)$, while the x measures whether the operation only increases the potential of diminishing $d(V, A \oplus A)$ in a subsequent operation.

Consider the possible operations that remain on a trajectory from T to U, i.e., if V_1 is transformed into V_2 by the operation, then $d(T, V_2) = d(T, V_1) + 1$ and $d(V_2, U) = d(V_1, U) - 1$. We set $y = d(V_1, A_1 \oplus A_1) - d(V_2, A_2 \oplus A_2) + 1$, where A_1 and A_2 are solutions of the halving problem for V_1 and V_2, respectively.

In evaluating an operation changing T^*, such as $\{a, b\}, \{c, d\} \to \{b, d\}, \{a, c\}$, we consider the following eight pairs: $\{a, b\}, \{c, d\}, \{b, d\}, \{a, c\}, \{\bar{a}, \bar{b}\}, \{\bar{c}, \bar{d}\}, \{\bar{b}, \bar{d}\}, \{\bar{a}, \bar{c}\}$.

The operation would clearly seem advantageous for subsequent operations if $\{\bar{b}, \bar{d}\}$ and/or $\{\bar{a}, \bar{c}\}$ were in T^* and/or U^*. There are from zero to four advantageous possibilities. In addition, although one of $\{b, d\}, \{a, c\}$ must be in U^* for the operation not to veer from an optimal trajectory, it is not necessary that both of them be. There are zero or one advantageous possibilities. We count how many h of the total of five advantageous possibilities occur and set $x = h + 1$.

The score W is in the range $[1, 18]$. We calculate W_{T^*} in this way and W_{U^*} by considering operations changing U^* in the direction of T^*. Let $W_X = \max_{\text{all operations}} W_{X^*}$.

If $W_T \geq W_U$ and $W_T \geq 6$, we apply the highest score operation to T^*. Otherwise apply the highest score operation to U^*, as long as this $W_U > 1$. The results of this operation and any other having the same score are added as nodes to a search tree. (The search tree was initialized when $T^* = T$ and $U^* = U$.) When there are no more operations that can be applied, we continue to build the search tree at a higher node. Finally, the leaves of the search tree are examined to find the highest scoring genome to be V, the last common ancestor of T and U.

Using a range $W \in [1, 18]$ proves clearly better than simply choosing evaluating an operation according to whether it $y = 1$ or $y \neq 1$. For example, in simulations generated with $d(T, V) = 60, d(V, U) = 55, d(V, A \oplus A) = 24$, the average estimate $d(V, A \oplus A)$ using an 18-value scale was 29.8, an overestimate of 24%, compared to 31.7 with a two-value scale, an overestimate of 32%.

5.3   Speciation First

In Section 5.2, $d(T, V) + d(V, U)$ was fixed and the problem was to find the common ancestor V with the shortest history from the doubling event. We now consider the halving distances of T and U both to be fixed, and look for the particular unduplicated genomes, ancestral to T and U, that are closest together. Our Algorithm 1 simultaneously halves T and U, choosing the initial vertex within each of the supernatural graphs (henceforward SNGs) so as to maximize the number of gray edges in common in the two ancestral genomes being constructed.

Both this heuristic and the one in Section 5.2 are basically $O(m^3)$ to arrive at a single estimate. This, however, generally produces a locally optimal solution. This is improved by maintaining a search tree in association with each algorithm. Then the running time is controlled by how large a search tree is maintained in the quest for lower estimates.

6   Simulations

Simulations of the doubling first model (five chromosomes, number of markers $m = 200$, inversions to translocations proportion 5:3, random choice of chromosomes to be rearranged, random breakpoints on chromosomes) show that our algorithm accurately reconstructs the number ν of rearrangements (ten replications for each value of ν) between the doubling event and the speciation event, as long as this is not too large (Fig. 6, top). With a longer interval between doubling and speciation, the halving algorithm reconstructs the unduplicated ancestor too economically. This, however, is a function not of the number of rearrangements in the simulation, but of the number of markers. If the number of markers is doubled from 200 to 400, the inferred number of rearrangements is corrected, as indicated by the square dot in the figure.

Simulations of the speciation first model ($m = 400$) show that while the genome halving distances accurately estimate the number of rearrangements between doubled ancestor and doubled descendant in the simulation (data not shown here), the estimated unduplicated ancestors are further apart than the genomes actually generated in the simulation (Fig. 6, bottom). This bias increases dramatically as a function, not of the distance itself, but of the amount of rearrangement these ancestors incur to produce the observed doubling descendant. When this "age" is 20, 50 and 80 rearrangements, the bias in the distance between the ancestors increases from 4 to 18 to 37, respectively. This reflects the severely non-unique result of the halving algorithm, which our algorithm attenuates by forcing the reconstructed doubled genomes to resemble each other as much as possible, but cannot eliminate, especially as the age of the doubling events recedes into the past.

Nonetheless, the superiority of our algorithm in constraining the two simultaneous halving processes to create ancestor genomes as close as possible, in comparison with a search over all pairs in $\mathbf{S_T} \times \mathbf{S_U}$, the Cartesian product of the two complete sets of solutions of the halving algorithm, is clear in another experiment.

Algorithm 1

Construct σ_T and σ_U, the set of supernatural graphs for T and U, respectively.

Initialize $\sigma_T^{(1)}$ = the subset of SNGs with 2 black edges and $\sigma_T^{(0)} = \sigma_T \setminus \sigma_T^{(1)}$

Initialize $\sigma_U^{(1)}$ = the subset of SNGs with 2 black edges and $\sigma_U^{(0)} = \sigma_U \setminus \sigma_U^{(1)}$

Step1: Order σ_T **and** σ_U

while there remain SNGs in $\sigma_T^{(0)}$ or SNGs in $\sigma_U^{(0)}$

> **while** there remain SNGs in $\sigma_T^{(0)}$ and either $\sigma_U^{(0)}$ is empty or the number of black edges in $\sigma_T^{(1)}$ is no more than in $\sigma_U^{(1)}$, we find a SNG in $\sigma_T^{(0)}$, to move from $\sigma_T^{(0)}$ to $\sigma_T^{(1)}$, as follows:
>
> > **for each** SNG s in $\sigma_T^{(0)}$, to count the maximum possible number of gray edges it could have in common with SNGs in $\sigma_U^{(1)}$:
> >
> > > **for** $i = 1, \cdots, |\sigma_U^{(1)}|$, if SNG s has k_i vertices in common with t_i, the i-th SNG in $\sigma_U^{(1)}$, the maximum number of gray edges they have in common is $[\frac{k_i}{2}]$.
> > >
> > > Then the score of s is $\sum_{i,\cdots,|\sigma_U^{(1)}|} [\frac{k_i}{2}]$.
> >
> > We add the highest scoring s to $\sigma_T^{(1)}$.
>
> **end while**
>
> **while** there remain SNGs in $\sigma_U^{(0)}$, and either $\sigma_T^{(0)}$ is empty or the number of black edges in $\sigma_U^{(1)}$ is less than $\sigma_T^{(1)}$), we find a SNG in $\sigma_U^{(0)}$, to move from $\sigma_U^{(0)}$ to $\sigma_U^{(1)}$, in the analogous way as for T
>
> **end while**

end while

Step2: Adding gray edges to σ_T **and** σ_U

For the root node of the search tree, add gray edges to all 2-edge SNGs in σ_T and σ_U

while there remain SNGs in σ_T or σ_U without gray edges.

> **while** there remain SNGs in σ_T without gray edges and either all SNGs in σ_U have gray edges or the number of gray edges in σ_T is no more than the number of gray edges in σ_U, let s be the first SNG in σ_T (according to the order in which it was added to $\sigma_T^{(1)}$) that has no gray edges. If s has l black edges, then we have l ways to choose the first black edge in this s, and 2 choices for orienting this edge, $2l$ choices in all, after which the dedouble algorithm proceeds deterministically to add gray edges within the SNG s.
>
> We add nodes to the search tree representing all the choices (out of the $2l$) that maximize the number of gray edges in common with σ_U.
>
> **end while**
>
> **while** there remain SNGs in σ_U without gray edges either all SNGs in σ_T have gray edges or the number of gray edges in σ_U is less than the number of gray edges in σ_T, let s be the first SNG in σ_U (according to the order in which it was added to $\sigma_U^{(1)}$) that has no gray edges. We use the same process as with T to get the best orderings within s and the associated gray edges.
>
> **end while**

end while

Solutions to the genome halving can then be found by tracing backwards from any leaf in the search tree.

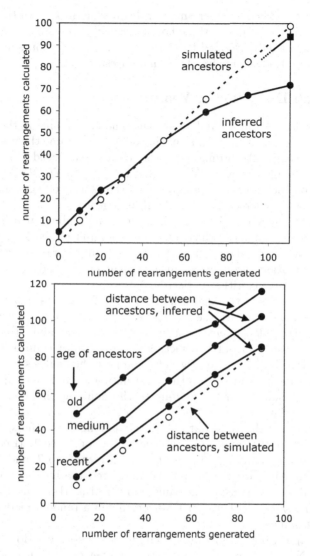

Fig. 6. Estimated distance: top, between doubling and speciation (age of ancestor=50), bottom, between unduplicated ancestors (ages: old=80, medium=50, young=20)

We set the initial number of markers to be 150, randomly assigned to 8 chromosomes. Then we carried out 45 random rearrangements to create one doubling ancestor and 38 independent rearrangements to create another. After tetraploidization formed two 300-marker genomes, we applied another 42 and 50 rearrangements, respectively, to create the modern doubling descendants. Then, using our knowledge of the ancestral genomes, we found that the distance between the two simulated ancestors was 75 and that the halving distances were 38 and 50, respectively. Using our speciation first algorithms on the two doubling descendants, we reached an inter-ancestor distance $d(A, B) = 84$ (instead of the simulated distance of 75) after three hours of calculation while the search of the Cartesian product only dropped to 87 (from 102) after 24 hours of calculation, involving almost 1,000,000 pairs of optimal ancestors.

7   Genome Doubling in Yeast

Wolfe and Shields [10] discovered an ancient genome doubling in the ancestry of *Saccharomyces cerevisiae* in 1997 after this organism became the first to have its genome sequenced [7]. According to [8], the recently sequenced *Candida glabrata* [4] shares this doubled ancestor. We extracted data from YGOB (Yeast Genome Browser) [3], on the orders and orientation of the exactly 600 genes identified as duplicates in both genomes, i.e., 300 duplicated genes.

We were able to obtain information from YGOB about which of the two duplicates in one genome is orthologous to which duplicate in the other genome. This is essential to the algorithm in Section 5.2. In general, we would have to infer this information through sequence comparison methods. This question is not pertinent to the algorithm in Section 5.3.

Though the results of the algorithm in Section 2 suggests that the theory in [8] is the most parsimonious, there is still enough uncertainty in yeast phylogenetics and enough independent occurrences of genome doubling, that it is worth comparing the results of our two methods to dispute or confirm the common doubled ancestor hypothesis. In Fig. 5 we compare the analysis in the left hand diagram with that in the right, on the yeast data and on data of approximately the same size generated first according to the doubling first model and then according to speciation first.

We first analyzed the yeast data using the doubling first and speciation first algorithms. The results appear in the centre row of Table 1. (Because of the asymmetry of the doubling first algorithm with respect to T and U, there are two sets of inferences for this case.) We then used the numbers of rearrangements inferred for yeast, using the same number of markers and chromosomes, to simulate the same number of rearrangements in a random model, both with doubling first and speciation first.

We then applied both algorithms, doubling first and speciation first, to both sets of data. Note first in Table 1 that the number of rearrangements inferred for the doubling first model using the doubling first algorithm is not exactly the same as that used to generate the data, and likewise for the speciation first case. This is normal, because the inference of rearrangements often is more economical than the rearrangements actually used.

The rows in Table 1 show that the doubling first analysis is better than the speciation first analysis (457 rearrangements versus 632) when the data are generated by doubling first, whereas the speciation first analysis is better (589 versus 604) when the data is generated with speciation first. The doubling first analysis clearly accounts better for the yeast data (505-521 versus 622), while the simulations assure that the biases in the two methods cannot be invoked, so our analysis confirms the hypothesis in [8].

Table 1. Doubling first (d.f) and speciation first (s.f.) analyses each produce a more parsimonious analysis of simulations produced by the corresponding model (d.f. or s.f., respectively). Averages of at least five simulations shown, but the effect holds for each simulation individually. The d.f. analysis gives a far better fit to the yeast data than s.f. Second yeast row reverses the roles of U and T in the algorithm.

analysis→	doubling first (d.f.)				speciation first (s.f.)			
data source↓	$d(T,V)$	$d(V,U)$	$d(V,A\oplus A)$	total	$d(T,A\oplus A)$	$d(A,B)$	$d(U,B\oplus B)$	total
sim by d.f.:	102	213	166	481				
inferred:	119	181	157	457	214	163	255	632
yeast:	92	245	168	505	193	179	250	622
	122	215	184	521				
sim by s.f.:					177	164	225	566
inferred:	146	354	104	604	164	228	197	589

8 Conclusions

Our previous work on integrating genome halving and other algorithms as a way of incorporating polyploids into rearrangement phylogeny used this software "off the shelf", searching all the many alternate outputs from one as inputs to the other. In the present paper we have avoided an exhaustive search strategy by intervening at the choice points in the genomic distance algorithm in the case of the doubling first problem and in the genome halving algorithm in the case of the speciation first problems. We have shown that these heuristics increase the efficiency of the search and to provide better upper bounds.

The main difficulty in this problem area remains the great multiplicity of solutions to the halving problem. Though this was only encountered here in the speciation first problem, leading to a overestimation of the inter-ancestor distance, it will also have to be dealt with in the doubling first scenario, when the inferred ancestor has to be integrated into a larger phylogenetic tree and compared to other doubled or unduplicated genomes, as in [11] and [9].

References

1. Bergeron, A., Mixtacki, J., Stoye, J.: A unifying view of genome rearrangements. In: Bücher, P., Moret, B.M.E. (eds.) WABI 2006. LNCS (LNBI), vol. 4175, pp. 163–173. Springer, Heidelberg (2006)
2. Bourque, G., Pevzner, P.: Genome-scale evolution: Reconstructing gene orders in the ancestral species. Genome Research 12, 26–36 (2002)
3. Byrne, K.P., Wolfe, K.H.: The Yeast Gene Order Browser: combining curated homology and syntenic context reveals gene fate in polyploid species. Genome Research 15, 1456–1461 (2005)
4. Dujon, B., Sherman, D., Fischer, G., et al.: Genome evolution in yeasts. Nature 430, 35–44 (2004)
5. El-Mabrouk, N., Sankoff, D.: Hybridization and genome rearrangement. In: Crochemore, M., Paterson, M.S. (eds.) Combinatorial Pattern Matching. LNCS, vol. 1645, pp. 78–87. Springer, Heidelberg (1999)

6. El-Mabrouk, N., Sankoff, D.: The reconstruction of doubled genomes. SIAM Journal on Computing 32, 754–792 (2003)
7. Goffeau, A., Barrell, B.G., Bussey, H., et al.: Life with 6000 genes. Science 275, 1051–1052 (1996)
8. Kurtzman, C.P., Robnett, C.J.: Phylogenetic relationships among yeasts of the Saccharomyces complex determined from multigene sequence analyses. FEMS Yeast Research 3, 417–432 (2003)
9. Sankoff, D., Zheng, C., Zhu, Q.: Polyploids, genome halving and phylogeny. Accepted for ISMB (2007)
10. Wolfe, K.H., Shields, D.C.: Molecular evidence for an ancient duplication of the entire yeast genome. Nature 387, 708–713 (1997)
11. Zheng, C., Zhu, Q., Sankoff, D.: Genome halving with an outgroup. Evolutionary Bioinformatics 2, 319–326 (2006)

A Rigorous Analysis of the Pattern of Intron Conservation Supports the *Coelomata* Clade of Animals

Jie Zheng, Igor B. Rogozin, Eugene V. Koonin*, and Teresa M. Przytycka*

National Center for Biotechnology Information, National Library of Medicine
National Institutes of Health, Bethesda, MD 20894, USA
{zhengj,rogozin,koonin,przytyck}@ncbi.nlm.nih.gov

Abstract. Many intron positions are conserved in varying subsets of eukaryotic genomes and, consequently, comprise a potentially informative class of phylogenetic characters. Roy and Gilbert developed a method of phylogenetic reconstruction using the patterns of intron presence-absence in eukaryotic genes and, applying this method to the analysis of animal phylogeny, obtained support for an Ecdysozoa clade ([1]). The critical assumption in the method was the independence of the rates of intron loss in different branches of the phylogenetic. Here, this assumption is refuted by showing that the branch-specific intron loss rates are strongly correlated. We show that different tree topologies are obtained, in each case with a significant statistical support, when different subsets of intron positions are analyzed. The analysis of the conserved intron positions supports the Coelomata topology, i.e., a clade comprised of arthropods and chordates, whereas the analysis of more variable intron positions favors the Ecdysozoa topology, i.e., a clade of arthropods and nematodes. We show, however, that the support for Ecdysozoa is fully explained by parallel loss of introns in nematodes and arthropods, a factor that does not contribute to the analysis of the conserved introns. The developed procedure for the identification and analysis of conserved introns and other characters with minimal or no homoplasy is expected to be useful for resolving many hard phylogenetic problems.

1   Introduction

Traditionally, molecular phylogenetics operates with alignments of homologous nucleotide or protein sequences that are used as the input for phylogenetic tree construction with one or another of the enormous variety of the available methods ([2]). Sequencing of numerous genomes from diverse taxa enabled the extension of phylogenetic analysis to the whole genome scale. Most often, this involves construction of phylogenetic trees from concatenated alignments of numerous genes or combination of numerous trees for individual gene sets into a supertree, but characters that can be properly denoted as genomic, such as

---

* Corresponding authors.

G. Tesler and D. Durand (Eds.): RECOMB-CG 2007, LNBI 4751, pp. 177–191, 2007.

gene composition, gene order, and protein domain combinations, have been employed as well ([3,4,5,6]). Methodologically, perhaps, the most promising category of genomic characters are rare genomic changes (RGCs) that represent, essentially, the genomic version of shared derived characters ("Hennigian" markers) ([7,8,9,10]). Insertion and deletion (gain and loss) of introns in protein-coding genes during the evolution of eukaryotes has been proposed as a promising class of RGCs ([8]). Indeed, positions in eukaryotic genes appear to be an attractive substrate for phylogenetic analysis because introns are extremely numerous, the positions of many but by no means all introns are conserved even between very distant eukaryotic taxa ([11,12]), and independent gain of introns in the same position in different lineages, which would lead to homoplasy, appears to be rare ([13]). Despite these potential advantages of intron positions as phylogenetic characters, there is a severe problem that complicates this approach, namely, the extensive parallel loss of introns in different lineages which leads to gross distortions of phylogenetic trees constructed on the basis of alignments of intron positions ([12]). To overcome this problem, Roy and Gilbert devised a statistical approach aimed at distinguishing between alternative phylogenetic hypotheses by comparing patterns of intron conservation ([1]).

Roy and Gilbert applied their method to the same set of alignments of intron positions that was previously analyzed by Rogozin et al. ([12]) and addressed one of the most debated, persistent problems in the large-scale animal phylogeny, namely, the controversy surrounding the Coelomata and Ecdysozoa topologies of the phylogenetic tree of animals. The "textbook" tree topology, originally stemming from comparative anatomy, includes a clade of animals that possess a true body cavity (coelomates, such as arthropods and chordates), whereas animals that have a pseudocoelome, such as nematodes, and those without a coelome, such as flatworms, occupy more basal positions in the tree (e.g., ([14,15]). The Coelomata topology appears "natural" from the viewpoint of straightforward and intuitive concept of the hierarchy of morphological and physiological complexity among animals, which is the main reason why this phylogeny had been accepted since the work of Haeckel ([16]). The first molecular phylogenetic analyses of 18S rRNA supported the Coelomata clade ([17,18]). However, in a seminal 1997 study, Lake and coworkers reported a new phylogenetic analysis of 18S rRNAs from a much larger set of animal species and arrived at an alternative tree topology that clustered arthropods and nematodes in a clade of molting animals termed Ecdysozoa ([19]).

The Ecdysozoa topology was recovered only when certain, apparently, slowly-evolving species of nematodes were included in the analyzed sample. Accordingly, it has been proposed that the coelomata topology is an artifact caused by long-branch attraction (LBA) ([19,20]) which is one of the most common artifacts of phylogenetic analysis ([21,22,23]). Specifically, the purported LBA has been attributed to the inclusion of fast-evolving species, such as nematodes of the genus Caenorhabditis. The ecdysozoan topology received additional support from the results of an independent phylogenetic analysis of 18S RNA ([24,25]), combined analysis of 18S and 28S rRNA sequences ([26]), and some protein phylogenies, such as those for Hox proteins ([27]). Furthermore, an apparent derived shared character of the Ecdysozoan clade has been identified, a distinct, multimeric form of β-thymosin ([28]).

Being compatible with the interpretation of molting as a fundamental developmental feature, the ecdysozoa topology has been rapidly and nearly universally accepted in the evo-devo community ([29,30,31,32]). However, phylogenetic analyses of multiple sets of orthologous proteins reopened the Coelomata-Ecdysozoa conundrum by consistently supporting the Coelomata topology ([33,34,35]). Both Blair et al. and Wolf et al. assessed the potential effect of branch length on the tree topology and concluded that the observed support for the Coelomata topology could not stem from LBA ([34,35]). Trees constructed by using non-sequence-based criteria, such as gene content and multidomain protein composition also supported Coelomata ([35]). Subsequently, the Coelomata topology received further support from several independent phylogenetic studies ([36,37,38,39]). In addition, the status of the multimeric β-thymosin as a derived shared character supporting the Ecdysozoa has been put into doubt as a result of the comparative analysis of recently sequenced genomes ([40]).

These reports have prompted further re-analyses including large-scale maximum-likelihood phylogenetic analysis of multiple genes from an extended range of animal species ([41,42,23]), putative derived characters, such as shared orthologs and domain combinations ([43]), and patterns of intron conservation in the aforementioned study of Roy and Gilbert ([1]). All these studies provided support for the Ecdysozoa topology suggesting, once again, that the coelomate topology was an LBA artefact, caused, largely, by inadequate taxon sampling and also, possibly, by the use of over-simplified models of sequence evolution ([41,23,44]).

Given the multiple reports in support for each of the alternative tree topologies, the Coelomata-Ecdysozoa dilemma is often considered unresolved, and accordingly, the metazoan tree is presented as a multifurcation ([45,46,47]). Very recently, we reexamined the problem using a new class of RGCs that include lineage specific replacements of amino acids that are, otherwise, conserved in a broad range of taxa and require two or three nucleotide substitutions; this study provided strong support for the Coelomata topology ([48]).

In a different study, a graph-theoretical method for identifying unstable phylogenetic characters recently developed by one of us was applied to remove, from intron position data, those positions that were found to be prone to multiple intron losses or independent gains. Phylogenetic analysis of the remaining intron positions strongly supported the Coelomata topology ([49]).

These findings prompted us to re-examine, in detail, the use of intron position conservation to infer the animal phylogeny. The central assumption of the method developed by Roy et al. is that the probability of retention of a given intron in the given branch can be modeled by a memory-less Markov process ([1]). Specifically, the probability of retaining an intron along any tree branch is assumed to depend on the branch but not on the retention history of a given intron in other branches of the tree. Nguyen et al. relied on the same assumption in their maximum-likelihood analysis of evolution of intron positions and, similarly, obtained support for the Ecdysozoa topology ([50]). Here we refute this assumption and demonstrate that an intron is more likely to be retained along a particular branch if it is also retained in other branches. Thus, it appears that introns are inherently more stable during evolution in some positions than in others, in an obvious parallel with different sites in proteins. Although the observed dependence of the retention probability on intron conservation in different branches invalidates the argument by Roy et al., it does not necessarily mean that data on intron presence-absence could not be used to infer phylogenies.

First, we evaluated whether the topology of the resulting tree depends on the level of intron conservation by partitioning the patterns of intron presence-absence into conserved and variable subsets. The analysis of the conserved intron set strongly supports the Coelomata phylogeny, in a direct contradiction with the conclusions of Roy and Gilbert, whereas the analysis of the variable introns yields the Ecdysozoa topology. It seems plausible that variable characters (intron positions, in this case) produce an incorrect tree due to long branch attraction. We rigorously show that, in a test of phylogenetic hypotheses with 5 taxa, the outcome is unaffected by parallel losses for characters that are conserved in three or four taxa, but is dramatically biased by parallel losses for characters that are present in only two taxa. In the context of the present study, this means that, if the analysis of the conserved intron set correctly yields the Coelomata topology, then, the Ecdysozoa topology is observed with the variable set due to parallel intron losses (homoplasy). We also present two additional, independent, formal arguments in support of the Coelomata topology.

2   Materials and Methods

Data set. The data set analyzed here is an extension of the previously described, curated set of conserved eukaryotic genes in which intron positions were mapped onto protein sequence alignments ([12]). In addition to the originally analyzed 8 species, namely, Anopheles gambiae (Ag), Arabidopsis thaliana (At), Caenorhabidits elegans (Ce), Drosophila melanogaster (Dm), Homo sapiens (Hs), Plasmodium falciparum (Pf), Saccharomyces cerevisiae (Sc), and Schizosaccharomyces pombe (Sp), two intron-rich fungi, Aspergillus fumigatus (Af) and Cryptococcus neoformans (Cn), and one Apicomplexan, Theileria parva (Tp), were included. The analyzed data set consists of 585 genes.

Each intron position corresponds to a binary string, where each bit (0 or 1) corresponds to a species and indicates whether an intron is present or absent at the given position of that species. We call such binary strings "patterns". To merge multiple species into one group, the bits of the merged species were replaced by a single bit that was 1 if any of the replaced bits was 1, and 0 otherwise. In the analysis, each pattern is a 5-bit string, including 4 animals and one outgroup (which in some cases consisted of several species). In our analysis, we considered three outgroups: Out 1 - Arabidopsis (At), Out 2 - four fungal species (Af, Cn, Sp, Sc), and Out 3 - all non-animal species in the data set (At,

Af, Cn, Sp, Sc, Tp, Pf). In the main body of the paper, we present the results for the largest outgourp (Out 3) as it provides the most robust data. The results for the other outgroups are presented in the Supplementary Materials.

For brevity, we usually refer to an intron position as an "intron", and when there is an intron in that position of a species, it is said that the intron occurs in that species. An intron position is considered conserved in two or more species if the intron occurs between two aligning bases in the alignment of the coding sequences. An intron occurs in a group if it occurs in any species of that group. To be informative for phylogenetic analysis, an intron must occur in at least two groups. This intron set was partitioned into conserved and variable subsets. The former subset consists of introns that occur in at least three groups, and the latter subset consists of the introns that occur in only two groups. The numbers of introns in the conserved and variable subsets are shown in Table 1.

Intron retention rates. The notations for comparison of retention rates are adopted from ([1]). Let A stand for arthropods (Dm and Ag), D for deuterostomes (Hs), N for nematodes (Ce), and O for outgroup. Then, ANO is the number of introns present in A, N, and O but absent in D, and ADON is the number of introns present in all four taxa. The retention rate is the ratio of introns present in a taxon to the number of introns that are absent (in the analyzed positions). For example, on the Coelomata phylogeny, for introns present in D and O, ADO/DO is the ratio of the number of introns retained along the branch A to the number of introns lost; similarly, ADON/DON is the retention rate in branch A for the introns that are also present in N. The p-values are the probabilities that the two ratios are equal, calculated using Fisher's exact test.

Dollo parsimony analysis of intron conservation patterns. Phylogenetic analysis of intron conservation patterns was performed using the Dollo parsimony method ([51,52]) which was applied either to all introns, or separately to the conserved and variable subsets using the Dolpenny program of the Phylip package ([53]) with default settings. For this analysis, the arthropods were split into two taxa, Dm and Ag, because conserved introns (those that occur in in more than two taxa) cannot resolve phylogenies for four taxa.

Table 1. The conserved and variable introns in the analyzed gene set, depending on the outgroup

Outgroup	Total # of introns	# variable introns	# conserved introns	# genes containing at least one conserved intron
Out 1[1]	1372	939	433	216
Out 2[2]	1394	925	469	212
Out 3[3]	1745	1203	542	242

---

[1] Arabidopsis (At)

[2] Fungal species (Af, Cn, Sp, Sc)

[3] All non-animal species in the data set (At, Af, Cn, Sp, Sc, Tp, Pf)

The statistical significance of the results was assessed using the winning sites test ([54]) as follows. For each intron, the minimum numbers of losses in the Ecdysozoa tree (E) and the Coelomata tree (C) was calculated, allowing for one gain only (the assumption of Dollo parsimony). If $E = C$, the intron is uninformative with respect to the support for one or the other topology and is discarded from the analysis; if $E > C$, the given intron supports Coelomata, and if $E < C$, the intron supports Ecdysozoa. Since, as observed from the data, for any intron, $|E - C| \leq 1$, the numbers of introns that favor each topology reflect the signals of dollo parsimony. The null hypothesis is that the probability that an intron favors Coelomata or Ecdysozoa is equal to 0.5. Using the binomial distribution, the p-value of a topology was calculated as the probability that at least the observed number of introns favor that topology. It should be noted, for clarity, that introns that are shared by two sister species cannot provide support for any hypothesis and are uninformative.

3   Results

3.1   The Dependence of Intron Loss Rate in a Branch on Intron Conservation in Other Branches

The central assumption of Roy and Gilbert's method ([1]) is that the probability of retention of a given intron in one branch can be modeled by a memory-less Markov process. Specifically, the probability of retaining an intron along any tree branch is assumed to depend on the branch but not on the retention history of the intron in the given position in other branches of the tree. Here, we refute this hypothesis and demonstrate that an intron in a given position is more likely to be retained along a particular branch if it is also retained in other branches.

We tested the null hypothesis that intron retention rate at any given tree branch is independent on whether or not the intron in the given position is retained in other (independent) branches of the tree. The hypothesis was tested for each of the two alternative topologies of the animal tree, Coelomata and

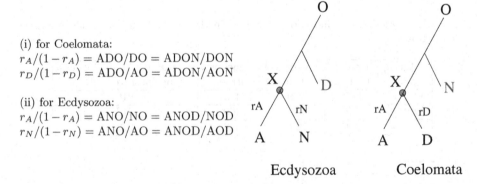

(i) for Coelomata:
$r_A/(1 - r_A) = \text{ADO/DO} = \text{ADON/DON}$
$r_D/(1 - r_D) = \text{ADO/AO} = \text{ADON/AON}$

(ii) for Ecdysozoa:
$r_A/(1 - r_A) = \text{ANO/NO} = \text{ANOD/NOD}$
$r_N/(1 - r_N) = \text{ANO/AO} = \text{ANOD/AOD}$

Fig. 1. Testing the independence of intron loss in different branches

Ecdysozoa. Let X be the common ancestor of Arthropods and Nematodes (for the Ecdysozoa tree) or Arthropods and Deuterostomes (for the Coelomata tree). Consider introns that are present in the out-group (O) and in at least one child of X. Under the assumption that the introns in the same position are orthologous (no convergent intron gain), such an intron must be present also in the node X but may or may not be present in D (for Ecdysozoa) or N (for Coelomata). We test the null hypothesis that the retention rates r_A, r_N and r_D (as appropriate for the corresponding tree topology) do not depend on whether or not the intron was retained in D (for Ecdysozoa) or N (for Coelomata) (Figure 1). If a lineage is represented by more than one species, the presence of the intron in any of these species implies that the intron is present at the root of the lineage. Then, testing the null hypothesis reduces to testing the equalities in Figure 1.

For the Ecdysozoa and Coelomata tree topologies, the null hypothesis is that the intron retention rates along the edges descending from the node X (r_A and either r_N or r_D, depending on the topology) is independent on whether the intron in the given position is also retained in D (for Ecdysozoa) or in N (for Coelomata).

The results for the largest outgroup (Out 3) are presented in Table 2. The results obtained with other outgroups are consistent with those in Table 2 (Supplementary Table S1 and Table S2). The null hypothesis consistently fails the Fisher's exact test, regardless of whether the Coelomata or the Ecdysozoa hypothesis is assumed. The difference in intron retention rates depending on the retention in other branches was not only statistically significant but, at least with some outgroups, quite dramatic. For example, for the Ecdysozoa topology and with the largest, group Out 3 (all non-animal species) as the outgroup, the retention rate of introns present in the deuterostomes was ∼0.43 whereas the corresponding retention rate for introns missing in the deuterostomes was only ∼0.15 (Table 2). Thus, the hypothesis that the probability of retaining an intron present in an ancestral node does not depend on the retention of this intron in other, independent branches of the tree is unequivocally rejected.

Table 2. Testing the dependence of intron retention on conservation in other branches, for Out 3

Ecdysozoa			
	Non-conserved in D	Conserved in D	p-value in Fisher
	ANO/NO	ANOD/NOD	test
$r_A/(1-r_A)$	11/85	100/146	8.01e-8
	ANO/AO	ANOD/AOD	
$r_N/(1-r_N)$	11/62	100/131	6.67e-6
Coelomata			
	Non-conserved in N	Conserved in N	p-value in Fisher
	ADO/DO	ADON/DON	test
$r_A/(1-r_A)$	131/711	100/146	1.08e-15
	ADO/AO	ADON/AON	
$r_D/(1-r_D)$	131/62	100/11	6.67e-6

3.2 Parsimony Analysis of Conserved Introns Supports the Coelomata Topology

Having shown that intron loss probability in a particular position critically depends on the state of that position in other branches, we reasoned that parallel loss that would distort the results of phylogenetic tree reconstruction is expected to be much more common among poorly conserved introns than among highly conserved ones. To examine the possible effect of intron conservation on the results of tree construction with the Dollo parsimony method, we analyzed trees for five species from the taxa of interest [Dm and Ag (Arthropods), Ce (Nematodes), Hs (Deuterostomes)], and an outgroup. We define variable introns as the introns that are present in two groups only. Conserved introns are defined as the introns that are present in at least tree groups. (For this analysis, 5 groups are required because, otherwise, the tree would be unresolved; therefore, it was necessary to treat the two Arthropod species as separate groups). The results for various choices of outgroup are shown in Supplementary Table S3. We call an intron position informative if the number of losses of the intron in this position differs between the Ecdysozoa and Coelomata tree topologies. The distribution of introns in informative positions provides support for one of the two trees.

We found that conserved introns consistently and highly significantly supported the Coelomata topology whereas variable introns yielded the Ecdysozoa tree. Given that variable introns substantially outnumber conserved introns, it is not surprising that the tree constructed using all introns is consistent with the Ecdysozoa topology (Supplementary Table S3). We repeated this analysis for other permutations of species replacing human by see urchin and one of the insects by bee and obtained consistent results.

3.3 Three Additional Tests to Resolve the Coelomata/Ecdysozoa Dilemma

We demonstrated that the Dollo parsimony tree constructed using conserved introns is consistent with the Coelomata hypothesis whereas the variable introns (and all introns, given that the variable introns comprise a substantial majority) supported the Ecdysozoa topology. Intuitively, it seems likely that the variable introns produce an erroneous result due to parallel losses in different branches. Nevertheless, we sought for specific, quantitative tests to distinguish between the two topologies. Three independent tests were developed, each based on a specific assumption about the evolutionary model. Since such models inevitably involve some level of simplification with respect to the true evolutionary scenario, we consider several increasingly more realistic assumptions. A corollary of the first test is that, if the Coelomata tree is correct, then, the fact that the tree obtained with variable (and with the full) set of introns displays the Ecdysozoa topology is a result of parallel intron losses. Taken together, the results of the tests described below not only lend strong support to the Coelomata topology but, through the above corollary, prove that the analysis of the unfiltered data leads to an incorrect tree due to parallel intron losses (a form of homoplasy).

Here we provide only the basic outline of the arguments; the complete description is provided in the Supplementary Materials.

Test 1. Assumptions: The argument is developed under the assumption of Dollo parsimony, i.e., irreversibility of intron loss. However, it also holds for the more general parsimony model where losses are treated as reversible (i.e., a loss and an independent gain of an intron occurring in the same position) as long as all character changes are weighted equally.

Argument outline: First, we rule out parallel loses (or independent gains) as a possible explanation of the tree topology obtained with conserved introns. Namely we show that none of the informative conserved introns could have undergone parallel losses, regardless of the tree topology (Coelomata or Ecdysozoa). In contrast, every informative variable intron has two parallel losses in one tree. Thus under the assumption that the tree obtained with conserved introns is correct, the inconsistency between this tree and the tree computed with variable introns is due to homoplasy.

This argument holds for any set of five taxa and any set of characters as long as the goal is to differentiate between two pre-defined tree topologies, such as Coelomata and Ecdysozoa. Thus, this approach resolves the discrepancy between the results obtained for a conserved set of characters and a variable set of characters in favor of the result obtained with the conserved characters as long, of course, as the number of such characters is sufficient to obtain statistically significant results.

Assumptions and simplifications: The argument holds independently of the assumption of character loss irreversibility (Dollo parsimony). However, it relies on the assumption that all changes have the same cost. This is an oversimplification, and a more realistic scenario should account for the different costs of intron loss in different branches. The second test we developed takes this into consideration.

Test 2. Assumptions: Intron loss is assumed to be irreversible. Assume that the cost of intron loss in the human branch and the C. elegans branch are different. (We can also associate weights with intron loss in other species but this is irrelevant for the argument). Let d_H and d_C be the costs of intron loss in the human branch and the C. elegans branch, respectively. Then, the phylogeny is constructed under a variant of the Dollo parsimony model where each loss is scored according to its assigned weight. Because, under parsimony, no a priori knowledge of the lengths of the internal edges is assumed, the costs of intron loss are assumed to be equal for all internal branches. Additionally, it is assumed that the ratio of the costs of loss between the variable and conserved sets of introns is approximately the same for the human and C. elegans branches. Since it is known that introns are lost on a massive scale in C.elegans but are highly conserved in humans ([55,56,57]), we assume $d_H/d_C > 1$.

Argument outline: We observe that the value of d_H/d_C affects the result of the Dollo parsimony reconstruction and we ask whether it is possible to set this

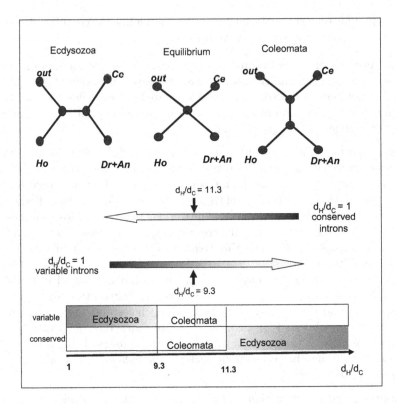

Fig. 2. The dependence of the tree topology on the d_H/d_C ratio. For the variable introns, $d_H/d_C = 1$ results in the Ecdysozoa tree and changes to the Coelomata tree for $d_H/d_C > 9.3$. For the conserved introns, $d_H/d_C = 1$ results in the Coelomata tree and changes to the Ecdysozoa tree for $d_H/d_C > 11.3$. For the interval $9.3 < d_H/d_C < 11.3$, the trees obtained from both sets agree on the Coelomata topology.

ratio so that conserved and variable introns produce the same tree. For each of the two sets of introns (the conserved and the variable ones), there exists an equilibrium value d_H/d_C where the Coelomata tree and the Ecdysozoa tree have the same cost (Figure 2). These equilibrium values can be computed directly from the frequencies of intron patterns (Table S4 and Table S5 in the Supplementary Material) as described in Supplementary Material. Using the data from the most robust outgroup, out3; we show that there exists an interval of the d_H/d_C ratio ($9.3 < d_H/d_C < 11.3$) for which both sets of introns (conserved and variable) produce the same tree and this tree has the Coelomata topology.

Possible shortcomings of the assumptions: In this argument, the cost of intron loss varies between species but the cost of intron loss in all internal branches is assumed to be the same. This oversimplification is removed in the next test.

Test 3. Assumption: For this test, intron loss is assumed to be irreversible but there are no constraints on the relation between the retention rates in different tree branches. It is expected that the retention rates. i.e., the ratio of the number of retained introns to the number of introns that were lost along a given branch, are higher for conserved than for variable introns.

Argument: We tested whether this expectation was violated under either of the two compared animal tree topologies. The retention rates were computed, under the Dollo parsimony assumption, for conserved and variable introns under the Coelomata and Ecdysozoa topologies (Supplementary Material, Table S6). For the Coelomata topology, the variable introns show a lower than or, approximately, the same retention rate as the conserved introns. By contrast, under the Ecdysozoa topology, the retention rate along the C. elegans branch was significantly greater for the variable introns than for the conserved introns ($p < 0.02$, one-sided Fisher test). This is the only significant deviation from the expectation regarding the retention rates that we observed, and it is seen in the Ecdysozoa tree. Thus, the test results support the Coelomata topology.

4   Discussion and Conclusions

Intron positions appear to be attractive candidates for the role of RGCs because eukaryotic genes contain numerous introns, thus, providing for statistically powerful phylogenetic tests and also because parallel gains of introns are rare ([13]). However, parallel intron losses in the same position are much more common and complicate phylogenetic analysis through the attraction of branches with high intron loss rates (a version of LBA). Roy and Gilbert as well as Nguyen et al. developed a method of phylogenetic reconstruction that overcame the problem of parallel losses but only under the assumption that the loss rates in different branches of the tree are independent; the application of this method to the animal phylogeny supported the existence of the Ecdysozoa clade ([1]). Here, we show that the independence assumption is invalid, i.e., the loss rates in different branches are strongly correlated. The outcome of phylogenetic analysis critically depends on the subset of intron positions that are used as the input. We show that, when exactly 5 taxa are used for phylogenetic analysis, for the introns that are conserved in three or four taxa, there are no parallel losses in informative positions, so the correct phylogeny is recovered so long as the assumption of the irreversibility of intron loss (no parallel gains) holds, at least, approximately, and the number of informative positions is sufficient to make the analysis statistically valid. In the specific case of animal phylogeny examined here, the analysis of such conserved introns strongly supports the Coelomata topology. By contrast, the analysis of variable introns (those represented in two taxa only) supports the Ecdysozoa topology, and because there are many more variable introns than conserved ones, the Ecdysozoa topology is recovered also when the entire set of introns is analyzed, in agreement with the observations of Roy and Gilbert ([1]). However, we proved that, if the topology obtained with the set of conserved introns (Coelomata) is valid, the recovery of the alternative topology (Ecdysozoa) is explained by parallel losses, in this case, in nematodes and insects.

With the results presented here, it appears that all tested RGCs including protein domain combinations ([35]), two-substitution replacements of highly conserved amino acids ([48]), and now intron positions support the Coelomata topology of the animal tree and reject the Ecdysozoa topology. However, some alternative analyses of RGCs ([43]) and several sequence-based phylogenetic studies employing extensive taxon sampling and sophisticated models of sequence evolution ([41,23,44]) support the Ecdysozoa and suggest that the Coelomata topology is an LBA artifact. This interpretation does not apply to our results obtained with the set of conserved introns; moreover, we specifically show that, when intron positions are used as phylogenetic characters, exactly the opposite is true, i.e., it is the Ecdysozoa topology that is produced as a result of LBA (extensive parallel loss of introns in arthropods and nematodes). The definitive resolution of the Coelomata-Ecdysozoa dilemma will require phylogenetic analysis of many more genomes from different branches of animals and reconciliation of the results obtained with various types of RGCs with the results of sequence-based phylogenetics.

The phylogenetic methodology described here, in principle, can be applied not only to introns but to any binary characters that meet the assumption of the irreversibility of losses that is required for the use of Dollo parsimony. Previously, attempts have been made to increase the accuracy of sequence-based phylogenies by limiting the analysis to slowly evolving positions in multiple alignments of protein and rRNA sequences, on the premise that such positions are the ones that are least prone to homoplasy ([58,59,60]). However, in this case, homoplasy could only be reduced by an uncertain amount, and there was the inevitable trade-off between the selection of increasingly conserved (presumably, increasingly homoplasy-free) positions and the loss of statistical power. The latter issue is pertinent also for the method described here, but at least in the present case study, the number of conserved introns, although a minority, was amply sufficient to allow an unequivocal discrimination between the two competing hypotheses. Moreover, applying this method to 5 taxa and using Dollo parsimony allows one not only to reduce but, actually, to eliminate a certain type of homoplasy. Therefore, the method is expected to be useful for resolving many hard phylogenetic problems.

Acknowledgments

This research was supported by the Intramural Research Program of the NIH, National Library of Medicine. The authors would like to thank Yuri I. Wolf for valuable discussions.

Supplementary Material

http://www.ncbi.nlm.nih.gov/CBBresearch/Przytycka/Coelomata/
intron07_som.doc

References

1. Roy, S.W., Gilbert, W.: Resolution of a deep animal divergence by the pattern of intron conservation. Proc. Natl. Acad. Sci. U S A 102, 4403–4408 (2005)
2. Felsenstein, J.: Inferring Phylogenies. Sinauer Associates, Sunderland, MA (2004)
3. Snel, B., Bork, P., Huynen, M.A.: Genome phylogeny based on gene content. Nat. Genet. 21, 108–110 (1999)
4. Wolf, Y.I., Rogozin, I.B., Grishin, N.V., Tatusov, R.L., Koonin, E.V.: Genome trees constructed using five different approaches suggest new major bacterial clades. BMC Evolutionary Biology. 1 (2001)
5. Wolf, Y.I., Rogozin, I.B., Grishin, N.V., Koonin, E.V.: Genome trees and the tree of life. Trends Genet. 18, 472–479 (2002)
6. Snel, B., Huynen, M.A., Dutilh, B.E.: Genome trees and the nature of genome evolution. Annu. Rev. Microbiol. 59, 191–209 (2005)
7. Rokas, A., Holland, P.W.: Rare genomic changes as a tool for phylogenetics. Trends in Ecology and Evolution 15, 454–459 (2000)
8. Nei, M., Kumar, S.: Molecular Evolution and Phylogenetics. Oxford Univ, Oxford (2001)
9. Delsuc, F., Brinkmann, H., Philippe, H.: Phylogenomics and the reconstruction of the tree of life. Nat. Rev. Genet. 6, 361–375 (2005)
10. Boore, J.L.: The use of genome-level characters for phylogenetic reconstruction. Trends Ecol. Evol. 21, 439–446 (2006)
11. Fedorov, A., Merican, A.F., Gilbert, W.: Large-scale comparison of intron positions among animal, plant, and fungal genes. Proc. Natl. Acad. Sci. U S A 99, 16128–16133 (2002)
12. Rogozin, I.B., Wolf, Y.I., Sorokin, A.V., Mirkin, B.G., Koonin, E.V.: Remarkable interkingdom conservation of intron positions and massive, lineage-specific intron loss and gain in eukaryotic evolution. Curr. Biol. 13, 1512–1517 (2003)
13. Sverdlov, A.V., Rogozin, I.B., Babenko, V.N., Koonin, E.V.: Conservation versus parallel gains in intron evolution. Nucleic Acids Res. 33, 1741–1748 (2005)
14. Brusca, R.C., Brusca, G.J.: Invertebrates. Sinauer Associates, Sunderland, Mass (1990)
15. Raff, R.A: The Shape of Life: Genes, Development, and the Evolution of Animal Form. University of Chicago Press, Chicago, IL (1996)
16. Haeckel, E.: Generelle Morphologie der Organismen. G.Reimer, Berlin (1866)
17. Field, K.G., Olsen, G.J., Lane, D.J., Giovannoni, S.J., Ghiselin, M.T., Raff, E.C., Pace, N.R., Raff, R.A.: Molecular phylogeny of the animal kingdom. Science 239, 748–753 (1988)
18. Turbeville, J.M., Pfeifer, D.M., Field, K.G., Raff, R.A.: The phylogenetic status of arthropods, as inferred from 18s rrna sequences. Mol. Biol. Evol. 8, 669–686 (1991)
19. Aguinaldo, A.M., Turbeville, J.M., Linford, L.S., Rivera, M.C., Garey, J.R., Raff, R.A., Lake, J.A.: Evidence for a clade of nematodes, arthropods and other moulting animals. Nature 387, 489–493 (1997)
20. Telford, M.J., Copley, R.R.: Animal phylogeny: fatal attraction. Curr. Biol. 15, 296–299 (2005)
21. Felsenstein, J.: Cases in which parsimony or compatibility methods will be positively misleading. Syst. Zool. 27, 401–410 (1978)
22. Reyes, A., Pesole, G., Saccone, C.: Long-branch attraction pheonomenon and the impact of among-site rate variation on rodent phylogeny. Gene 259, 177–187 (2000)

23. Philippe, H., Lartillot, N., Brinkmann, H.: Multigene analyses of bilaterian animals corroborate the monophyly of ecdysozoa, lophotrochozoa, and protostomia. Mol. Biol. Evol. 22, 1246–1253 (2005)

24. Giribet, G., Distel, D.L., Polz, M., Sterrer, W., Wheeler, W.C.: Triploblastic relationships with emphasis on the acoelomates and the position of gnathostomulida, cycliophora, plathelminthes, and chaetognatha: a combined approach of 18s rdna sequences and morphology. Syst. Biol. 49, 539–562 (2000)

25. Peterson, K.J., Eernisse, D.J.: Animal phylogeny and the ancestry of bilaterians: inferences from morphology and 18s rdna gene sequences. Evol. Dev. 3, 170–205 (2001)

26. Mallatt, J., Winchell, C.J.: Testing the new animal phylogeny: first use of combined large-subunit and small-subunit rrna gene sequences to classify the protostomes. Mol. Biol. Evol. 19, 289–301 (2002)

27. de Rosa, R., Grenier, J.K., Andreeva, T., Cook, C.E., Adoutte, A., Akam, M., Carroll, S.B., Balavoine, G.: Hox genes in brachiopods and priapulids and protostome evolution. Nature 399, 772–776 (1999)

28. Manuel, M., Kruse, M., Muller, W.E., Parco, Y.L.: The comparison of beta-thymosin homologues among metazoa supports an arthropod-nematode clade. J. Mol. Evol. 51, 378–381 (2000)

29. Adoutte, A., Balavoine, G., Lartillot, N., Lespinet, O., Prud'homme, B., de Rosa, R.: The new animal phylogeny: reliability and implications. Proc. Natl. Acad. Sci. U S A 97, 4453–4456 (2000)

30. Valentine, J.W., Collins, A.G.: The significance of moulting in ecdysozoan evolution. Evol. Dev. 2, 152–156 (2000)

31. Collins, A.G., Valentine, J.W.: Defining phyla: evolutionary pathways to metazoan body plans. Evol. Dev. 3, 432–442 (2001)

32. Telford, M.J., Budd, G.E.: The place of phylogeny and cladistics in evo-devo research. Int. J. Dev. Biol. 47, 479–490 (2003)

33. Mushegian, A.R., Garey, J.R., Martin, J., Liu, L.X.: Large-scale taxonomic profiling of eukaryotic model organisms: a comparison of orthologous proteins encoded by the human, fly, nematode, and yeast genomes. Genome Res. 8, 590–598 (1998)

34. Blair, J.E., Ikeo, K., Gojobori, T., Hedges, S.B.: The evolutionary position of nematodes. BMC Evol. Biol. 2(7) (2002)

35. Wolf, Y.I., Rogozin, I.B., Koonin, E.V.: Coelomata and not ecdysozoa: evidence from genome-wide phylogenetic analysis. Genome Res. 14, 29–36 (2004)

36. Stuart, G.W., Berry, M.W.: An svd-based comparison of nine whole eukaryotic genomes supports a coelomate rather than ecdysozoan lineage. BMC Bioinformatics 5, 204 (2004)

37. Philip, G.K., Creevey, C.J., McInerney, J.O.: The opisthokonta and the ecdysozoa may not be clades: stronger support for the grouping of plant and animal than for animal and fungi and stronger support for the coelomata than ecdysozoa. Mol. Biol. Evol. 22, 1175–1184 (2005)

38. Zdobnov, E.M., von Mering, C., Letunic, I., Bork, P.: Consistency of genome-based methods in measuring metazoan evolution. FEBS Lett. 579, 3355–3361 (2005)

39. Ciccarelli, F.D., Doerks, T., von Mering, C., Creevey, C.J., Snel, B., Bork, P.: Toward automatic reconstruction of a highly resolved tree of life. Science 311, 1283–1287 (2006)

40. Telford, M.J.: The multimeric beta-thymosin found in nematodes and arthropods is not a synapomorphy of the ecdysozoa. Evol. Dev. 6, 90–94 (2004)

41. Brinkmann, H., van der Giezen, M., Zhou, Y., de Raucourt, G.P., Philippe, H.: An empirical assessment of long-branch attraction artefacts in deep eukaryotic phylogenomics. Syst. Biol. 54, 743–757 (2005)
42. Dopazo, H., Dopazo, J.: Genome-scale evidence of the nematode-arthropod clade. Genome Biol 6(5), R41 (2005)
43. Copley, R.R., Aloy, P., Russell, R.B., Telford, M.J.: Systematic searches for molecular synapomorphies in model metazoan genomes give some support for ecdysozoa after accounting for the idiosyncrasies of caenorhabditis elegans. Evol. Dev. 6, 164–169 (2004)
44. Lartillot, N., Brinkmann, H., Philippe, H.: Suppression of long-branch attraction artefacts in the animal phylogeny using a site-heterogeneous model. BMC Evol. Biol. Suppl. 1 7, S4 (2007)
45. Hedges, S.B.: The origin and evolution of model organisms. Nat. Rev. Genet. 3, 838–849 (2002)
46. Telford, M.J.: Animal phylogeny: back to the coelomata? Curr. Biol. 14, R274–276 (2004)
47. Jones, M., Blaxter, M.: Evolutionary biology: animal roots and shoots. Nature 434, 1076–1077 (2005)
48. Rogozin, I.B., Wolf, Y.I., Carmel, L., Koonin, E.V.: Ecdysozoan clade rejected by genome-wide analysis of rare amino acid replacements. Mol. Biol. Evol. 24, 1080–1090 (2007)
49. Przytycka, T.M.: An important connection between network motifs and parsimony models. In: Apostolico, A., Guerra, C., Istrail, S., Pevzner, P., Waterman, M. (eds.) RECOMB 2006. LNCS (LNBI), vol. 3909, pp. 321–335. Springer, Heidelberg (2006)
50. Nguyen, H.D, Yoshihama, M., Kenmochi, N.: New maximum likelihood estimators for eukaryotic intron evolution. PLoS Comput. Biol. 1(7), 79 (2005)
51. Farris, J.S.: Phylogenetic analysis under dollo's law. Syst. Zool. 26, 77–88 (1977)
52. Rogozin, I.B., Babenko, V.N., Wolf, Y.I., Koonin, E.V.: Dollo parsimony and reconstruction of genome evolution. In: Albert, V.A. (ed.) Parsimony, Phylogeny, and Genomics, pp. 190–200. Oxford University Press, Oxford (2005)
53. Felsenstein, J.: Inferring phylogenies from protein sequences by parsimony, distance, and likelihood methods. Methods Enzymol 266, 418–427 (1996)
54. Prager, E.M., Wilson, A.C.: Ancient origin of lactalbumin from lysozyme: analysis of dna and amino acid sequences. J. Mol. Evol. 27, 326–335 (1988)
55. Fedorov, A., Roy, S., Fedorova, L., Gilbert, W.: Mystery of intron gain. Genome Res. 13, 2236–2241 (2003)
56. Roy, S.W., Penny, D.: Smoke without fire: most reported cases of intron gain in nematodes instead reflect intron losses. Mol. Biol. Evol. 23, 2259–2262 (2006)
57. Carmel, L., Wolf, Y.I., Rogozin, I.B., Koonin, E.V.: Three distinct modes of intron dynamics in the evolution of eukaryotes. Genome Res. (in press, 2007)
58. Brinkmann, H., Philippe, H.: Archaea sister group of bacteria? indications from tree reconstruction artifacts in ancient phylogenies. Mol. Biol. Evol. 16, 817–825 (1999)
59. Philippe, H., Germot, A., Moreira, D.: The new phylogeny of eukaryotes. Curr. Opin. Genet. Dev. 10, 596–601 (2000)
60. Brochier, C., Philippe, H.: Phylogeny: a non-hyperthermophilic ancestor for bacteria. Nature 417, 244 (2002)

Author Index

Lecture Notes in Bioinformatics

Vol. 3380: C. Priami (Ed.), Transactions on Computational Systems Biology I. IX, 111 pages. 2005.

Vol. 3370: A. Konagaya, K. Satou (Eds.), Grid Computing in Life Science. X, 188 pages. 2005.

Vol. 3318: E. Eskin, C. Workman (Eds.), Regulatory Genomics. VII, 115 pages. 2005.

Vol. 3240: I. Jonassen, J. Kim (Eds.), Algorithms in Bioinformatics. IX, 476 pages. 2004.

Vol. 3082: V. Danos, V. Schachter (Eds.), Computational Methods in Systems Biology. IX, 280 pages. 2005.

Vol. 2994: E. Rahm (Ed.), Data Integration in the Life Sciences. X, 221 pages. 2004.

Vol. 2983: S. Istrail, M.S. Waterman, A. Clark (Eds.), Computational Methods for SNPs and Haplotype Inference. IX, 153 pages. 2004.

Vol. 2812: G. Benson, R.D.M. Page (Eds.), Algorithms in Bioinformatics. X, 528 pages. 2003.

Vol. 2666: C. Guerra, S. Istrail (Eds.), Mathematical Methods for Protein Structure Analysis and Design. XI, 157 pages. 2003.